AF425618

Generative Art

Amir Husain (@amirhusain_tx)

ISBN-13: 979-8-9884751-1-8 (Hardcover)
ISBN-13: 979-8-9884751-3-2 (Paperback)

For Asas, Murtaza, and Hyder.
May their curiosity forever grow,
and with it, their creativity.

Contents

Introduction

Generative art is a captivating form of artistic expression that leverages algorithms, mathematics, and technology to create stunning visuals, novel effects, and deep experiences. These artistic works are not merely the product of human imagination, but the result of a collaborative process between the artist and the machine.

I've been fascinated with generative applications for a long time. Computers can so easily iterate over a large number of possibilities and find beautiful pictures, interesting solutions and novel structures that would take humans a very long time to chance upon.

Generative art using computers has been produced for many years now. Fractals, for example, were incredibly popular in the 1980s and 90s, and a large number of programs for generating the Mandelbrot and Julia sets proliferated on home computers. And genre-creating bands like Kraftwerk used simple, repetitive waveforms generated with electronics, overlaid to produce electronic music the likes of which hadn't been heard before. Most of the time, however, these programs were not particularly easy for non-technical users to employ. So, the interest in Generative art remained niche, shared mainly by some artistically inclined programmers and technically inclined artists. A small universe!

In this book, I want to share my love of Generative art with you by walking you through Generative Designer, a small but highly customizable Generative art program I have written. Generative Designer has been designed to offer an effective introduction to the world of computational art for artists and enthusiasts. If you are not a programmer or not comfortable modifying code to create art, you can simply use Generative Designer by manipulating its easy-to-use interface for creating mesmerizing images.

In the pages ahead, we will explore the world of Generative art and examine how Generative Designer can be used to create some interesting and, at times, even beautiful images.

By following along with the examples provided, you will gain a deeper understanding of the principles behind Generative art and learn how to harness the power of Generative Designer to bring your unique visions to life.

Chapter 1

The Origin of the Serendipitous Circles Algorithm

Where did the idea for Generative Designer come from? Well, from my techno-archaeology hobby, actually!

The algorithm implemented in the Generative Designer program is inspired by an article I found in my old collection of BYTE magazines. BYTE magazine was an absolutely iconic computer magazine published from 1975 to 1998. It was well-known for its in-depth coverage of computers and software and was particularly popular among hobbyists and computer enthusiasts. I remember impatiently waiting for the new BYTE to show up at my local bookstore. It was hard to wait because each month the magazine provided detailed technical articles, how-tos, news and reviews. It was truly a source of inspiration for many computer programmers and designers. Since this is a discussion of computers and art, BYTE was also noted for its high-quality illustrations and diagrams. The magazine's cover art is still admired and printed on T-shirts and posters. Unfortunately, the magazine ceased publication in 1998. My collection of original BYTE magazines is one of the most treasured parts of my library, and I often find really useful code and ideas from long, long ago. Technology may have changed, but many of the ideas captured on the pages of BYTE remain evergreen.

One such timeless article led me on an exploration which is culminating with this book. Titled "Serendipitous Circles Explored", and written by Eduardo Kellerman, it discussed a simple algorithm for generating computer art. BYTE first discussed this algorithm in August 1977 in a technical article written by D.J. Anderson and W.F. Galway.

The "Serendipitous Circles" algorithm is surprisingly flexible, and for its compact size, offers a fascinating variety of abstract outputs. The algorithm computes a series of (X, Y) pairs and displays them on a graphics device. Each (X, Y) pair is computed from the preceding pair using two equations, one for X and one for Y. The only requirement is to supply an initial (X, Y) pair and iteration, or the repeated application of the functions that derive the next coordinates, does the rest.

The circles and ovals that emerge via this algorithm are "reflected" in kaleidoscopic fashion to create symmetry across the screen. The original algorithm was used to plot images on 1970s era terminals and oscilloscope displays. Thankfully, with web browsers, JavaScript and multi-color high-resolution displays, we have something far more modern with which to implement this algorithm.

What I found fascinating and surprising as I played with basic implementations of the algorithm described in "Serendipitous Circles" was how diverse its range of output images can be. In a few lines of code, it provides a simple and effective method for generating computer art which can be customized via a variety of parameters. By experimenting with different equations and overlapping the results of using several (X, Y) starting pairs, unique and interesting patterns emerge. I've experimented with many here, and the Generative Designer program I cover in this book allows several parameters to be customized by the end user. However, there is even more scope for modification and improvement, and I will discuss some of what can be done in the future.

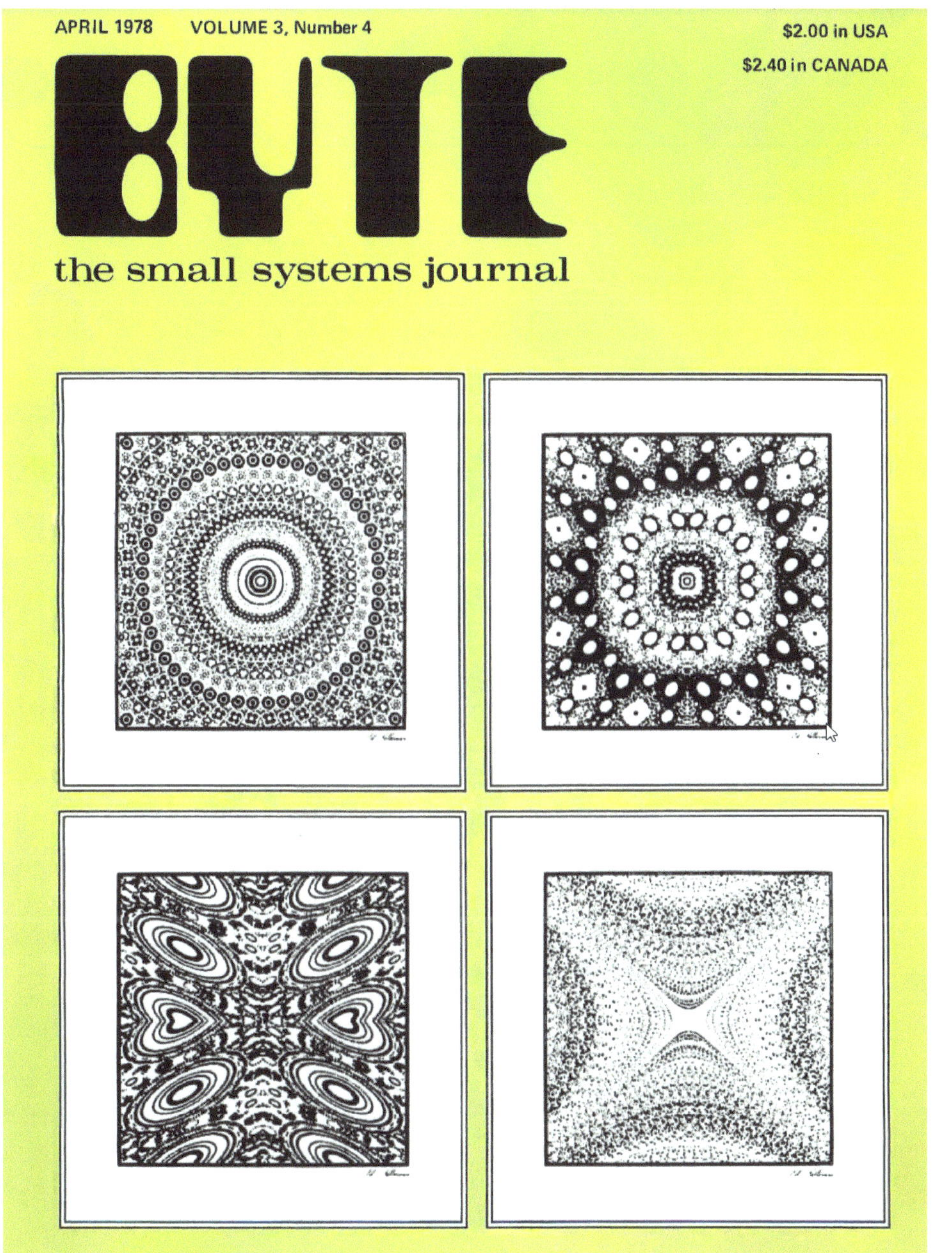

Figure 1.1: The BYTE April 1978 cover.

Chapter 2

The Generative Designer

2.1 About the Program

The instructive value of the Generative Designer program lies in its ability to produce complex, absorbing images through a combination of simple user inputs and the creative "Serendipitous Circles" algorithm. The program offers a set of controls that allow users to adjust various parameters such as color palettes, line widths, and rotation angles. By tweaking these settings, you can experiment with an almost infinite number of combinations to create images that are truly novel.

Generative Designer utilizes a number of techniques from the world of computer graphics and mathematics such as function plotting, translations, and image rotations around a center. The core of the program is built using p5.js, a powerful JavaScript library that simplifies the process of creating graphics and interactive content for the web. p5.js is incredibly popular with artists and is often used to write Generative art algorithms. Using this library, Generative Designer can efficiently generate high-quality images that can be easily shared and displayed on various platforms.

Because the algorithms are implemented in JavaScript and designed to run within your web browser, Generative Designer uses your local computer resources to create images. If you are trying to generate complex images with lots of connected lines and a high number of iterations, you might need a reasonably modern computer. The current implementation of Generative Designer does not support rendering on phones, but future versions may add support for mobile devices.

As you progress through these pages and explore the various examples of Generative art, you will hopefully gain a deeper appreciation for not only the art of Generative Designer, but the art of the possible with generative design. Together, we will walk through the options and flexibility that Generative Designer has to offer. Whether you are an experienced artist or simply someone with a keen interest in visual arts, I hope this book will provide you with ideas and tools with which you can understand and pursue projects in Generative art and design.

2.2   How to access Generative Designer

As of April 2023, there is a running version of Generative Designer deployed at the following web address `http://openworks.cc/gd`. You can point your computer browser to this address and try the application out for yourself. No registration is required, no telemetry of any kind regarding the settings you use is stored, nor is any of the generated data stored on the remote server. In fact, all computation is done locally on your own machine using the JavaScript virtual machine (VM) inside your own browser.

You can save images created by Generative Designer by simply right-clicking the generated image and saving it to disk.

2.3　A Tour of the Application

Now that we know what Generative Designer is all about, let's go on a quick tour of the application and see how the various input settings available to us can affect design outcomes.

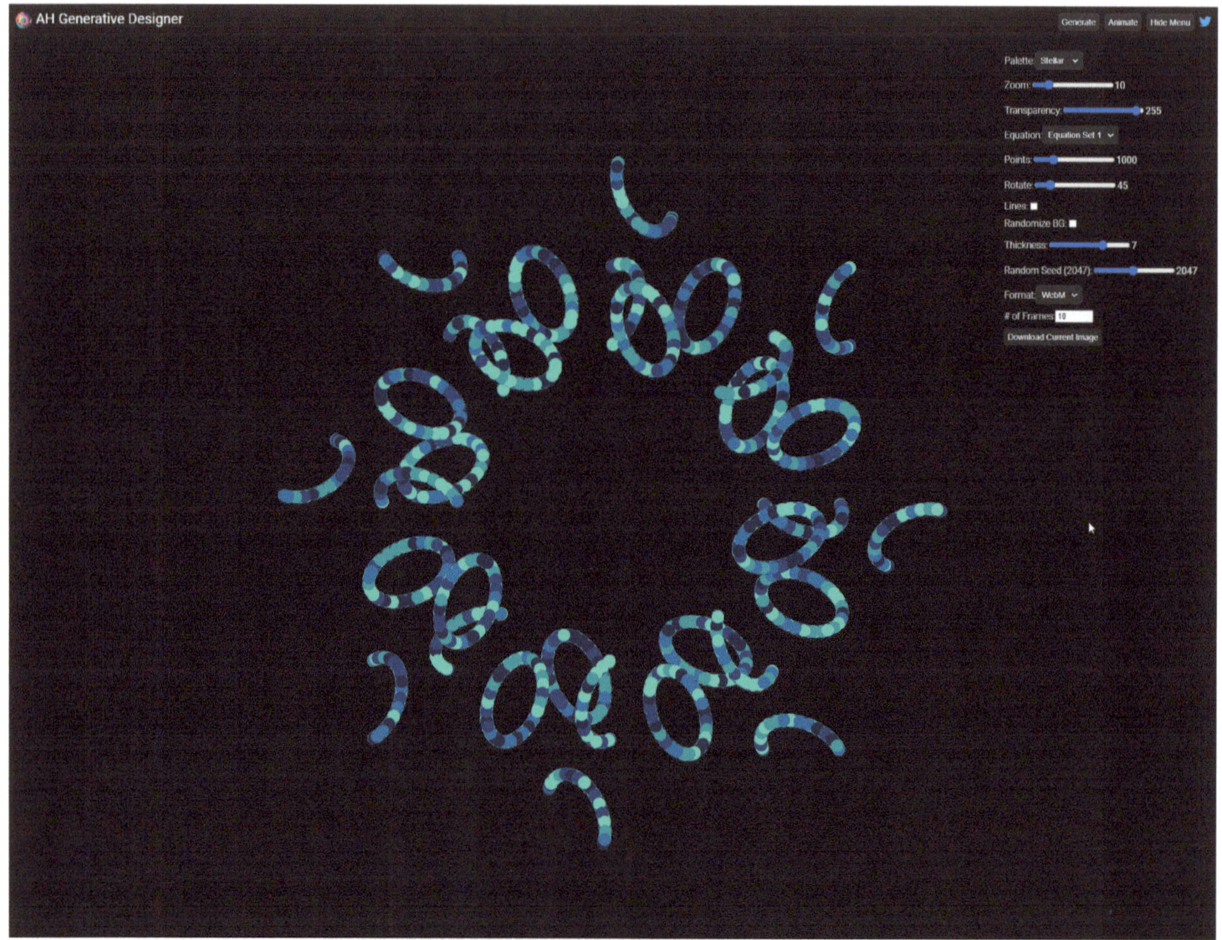

Figure 2.1: The web-based user interface of Generative Designer. The controls are simple and available at top right. Clicking on the Generate button creates an image. Clicking on Animate created a video of random variations of the same image input parameters. And clicking on Show Menu displays all the settings available to modify the image type and style.

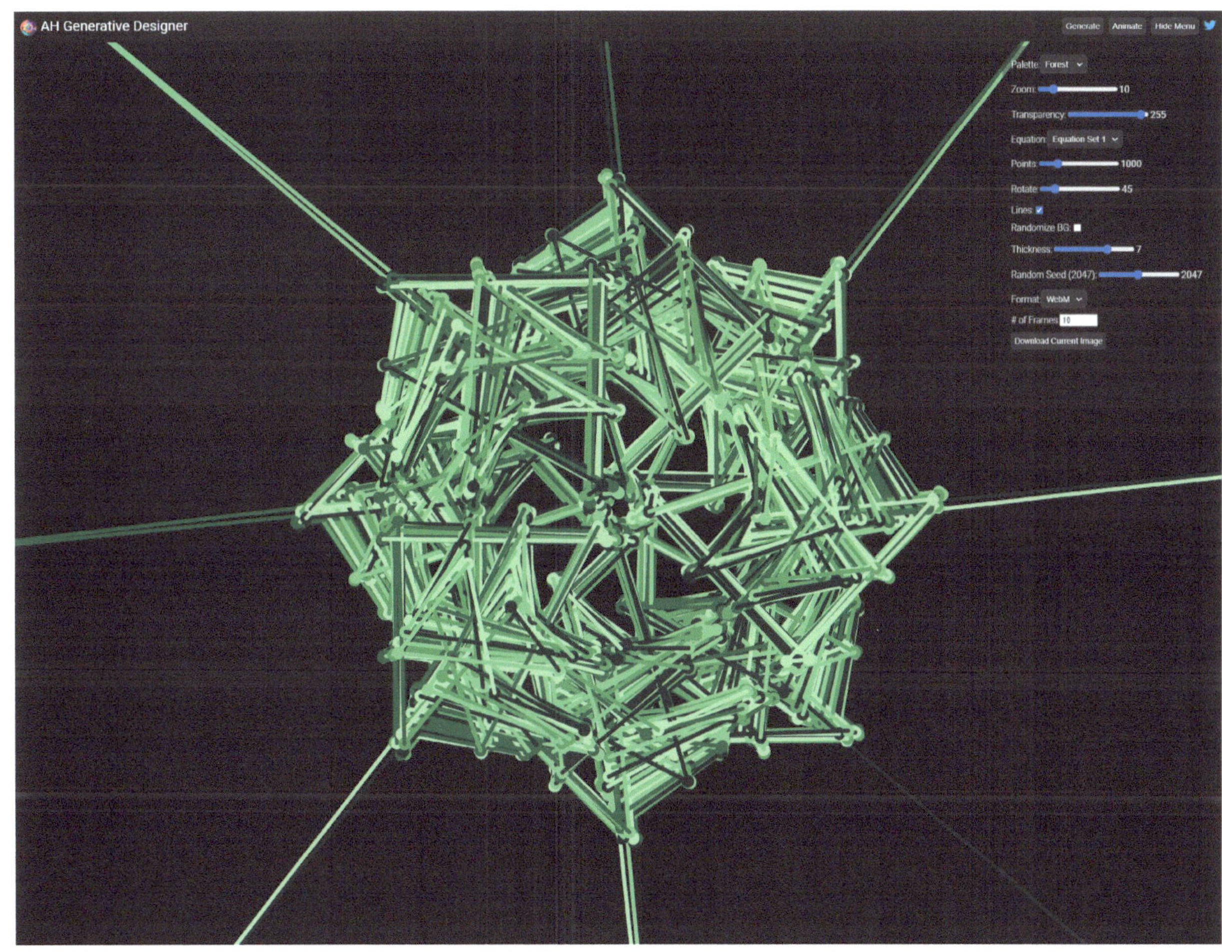

Figure 2.2: With the Lines checkbox turned on, you can connect all plotted points.

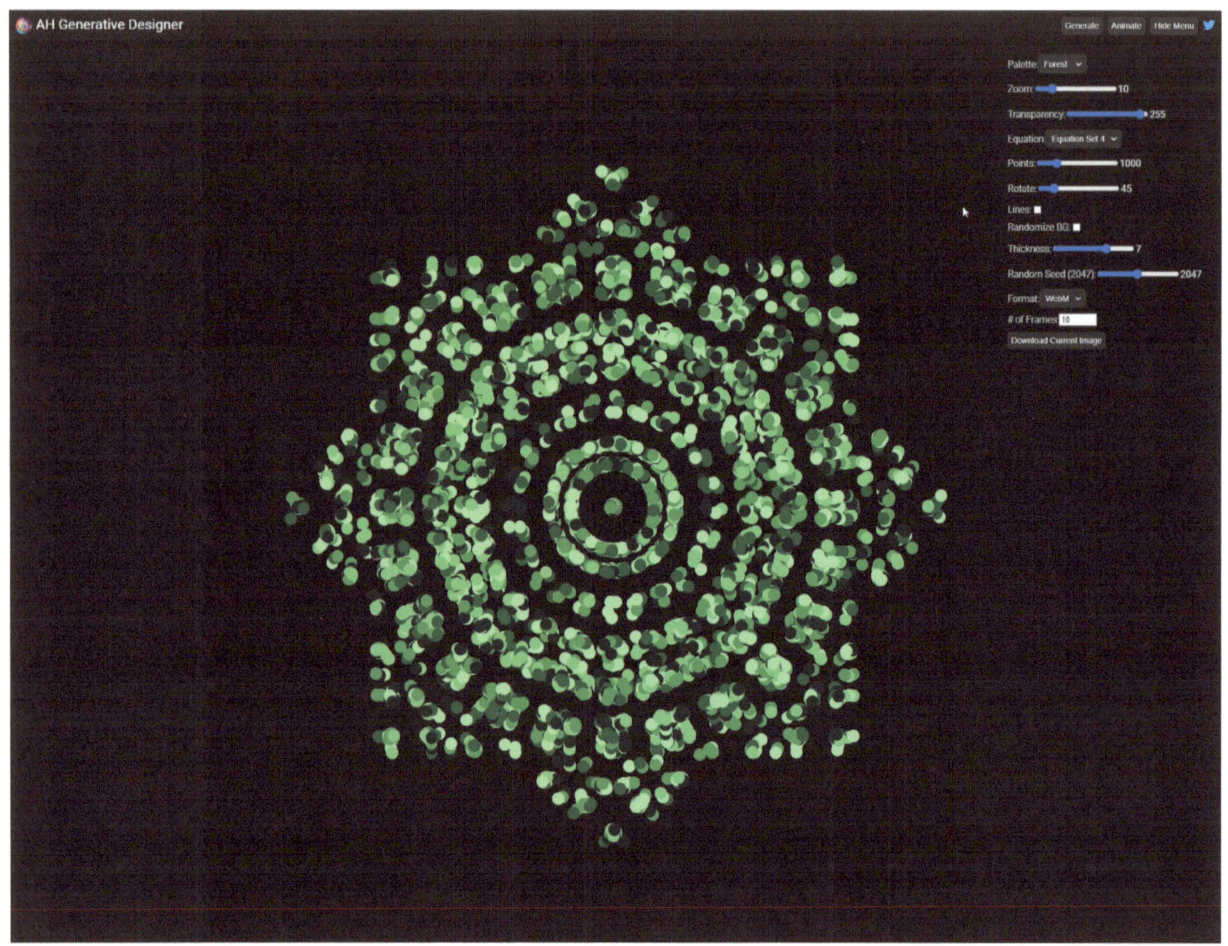

Figure 2.3: With Lines turned off, the underlying plot becomes apparent.

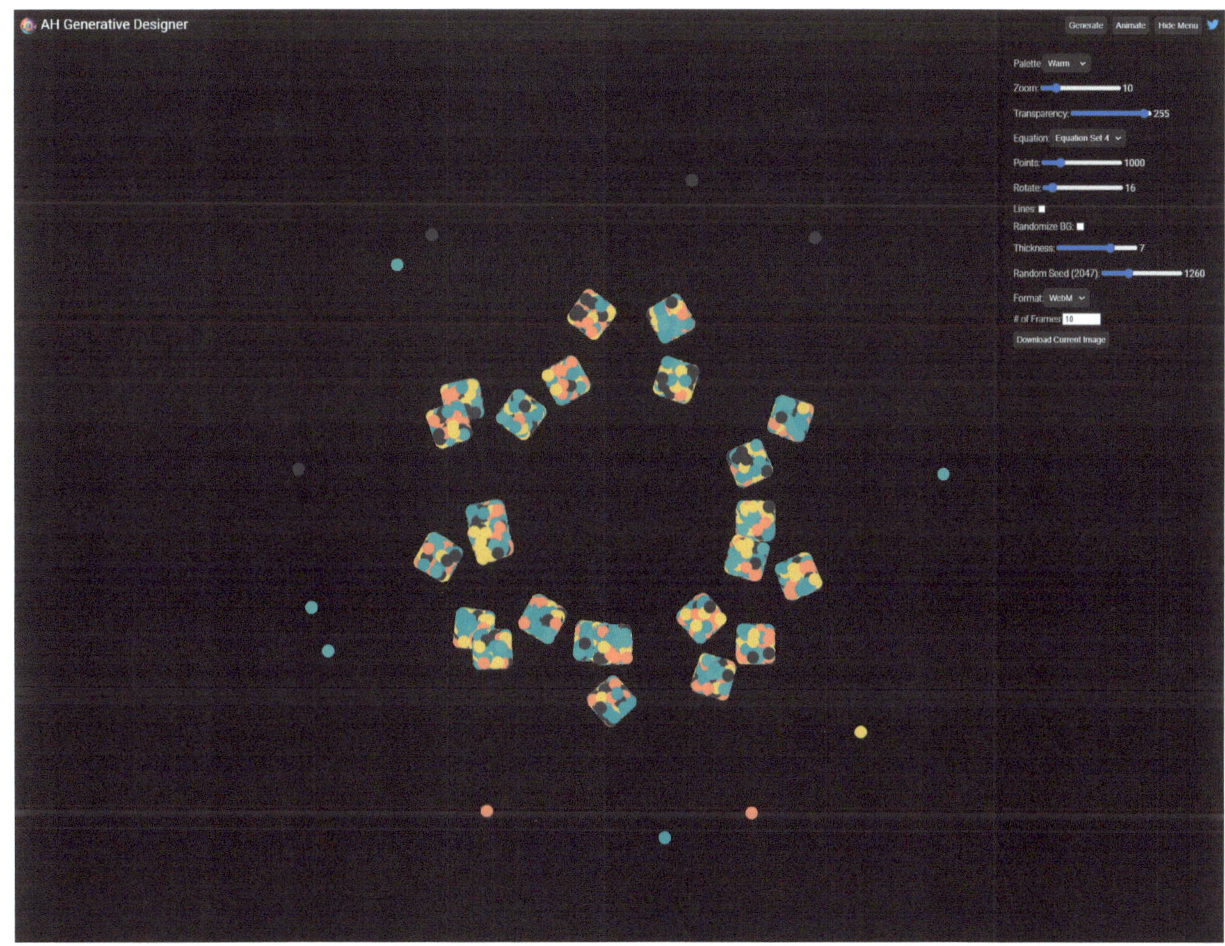

Figure 2.4: Adjusting the Rotate slider replicates parts of the image, but without perfect symmetry.

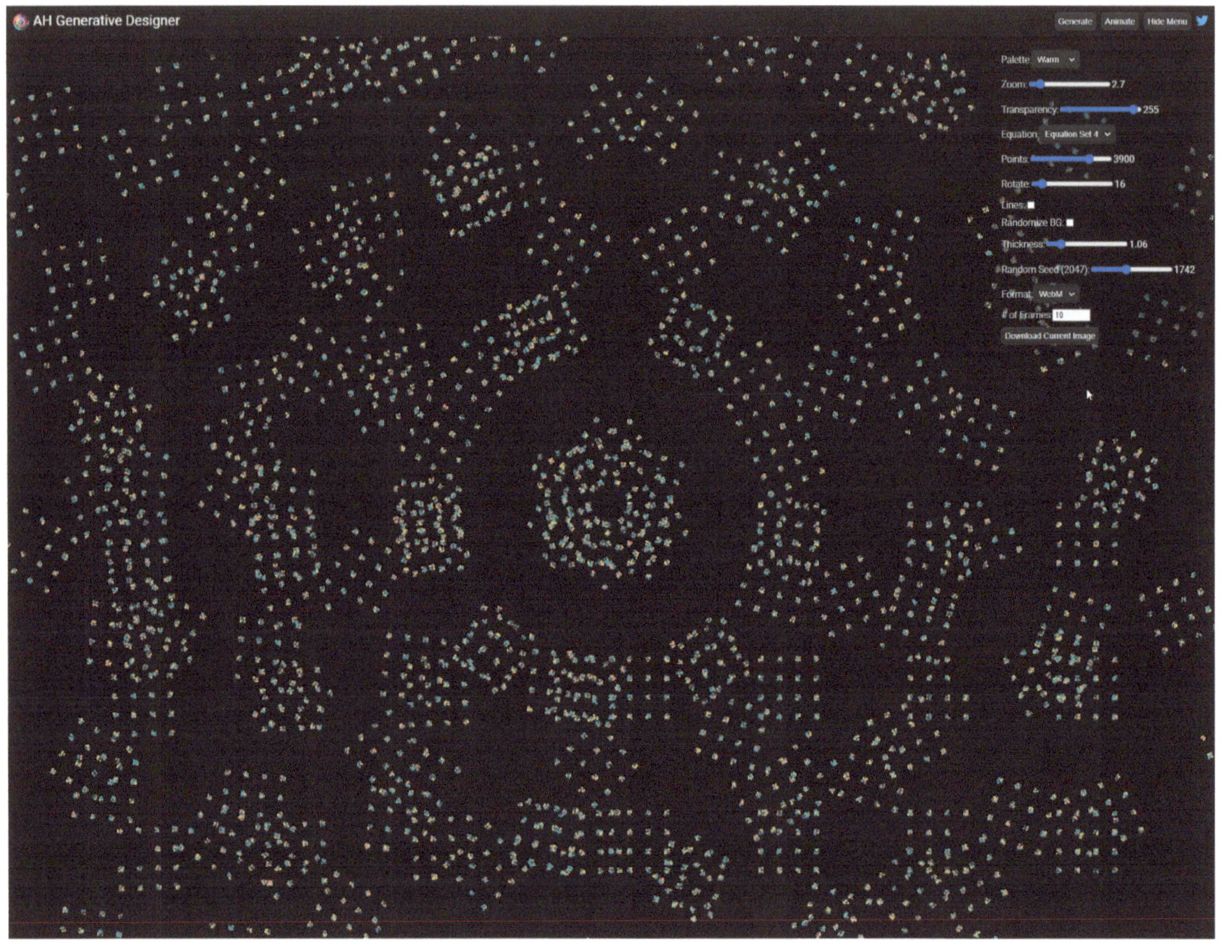

Figure 2.5: Adjusting the Thickness of the "brush" drastically changes the mood and look of a piece.

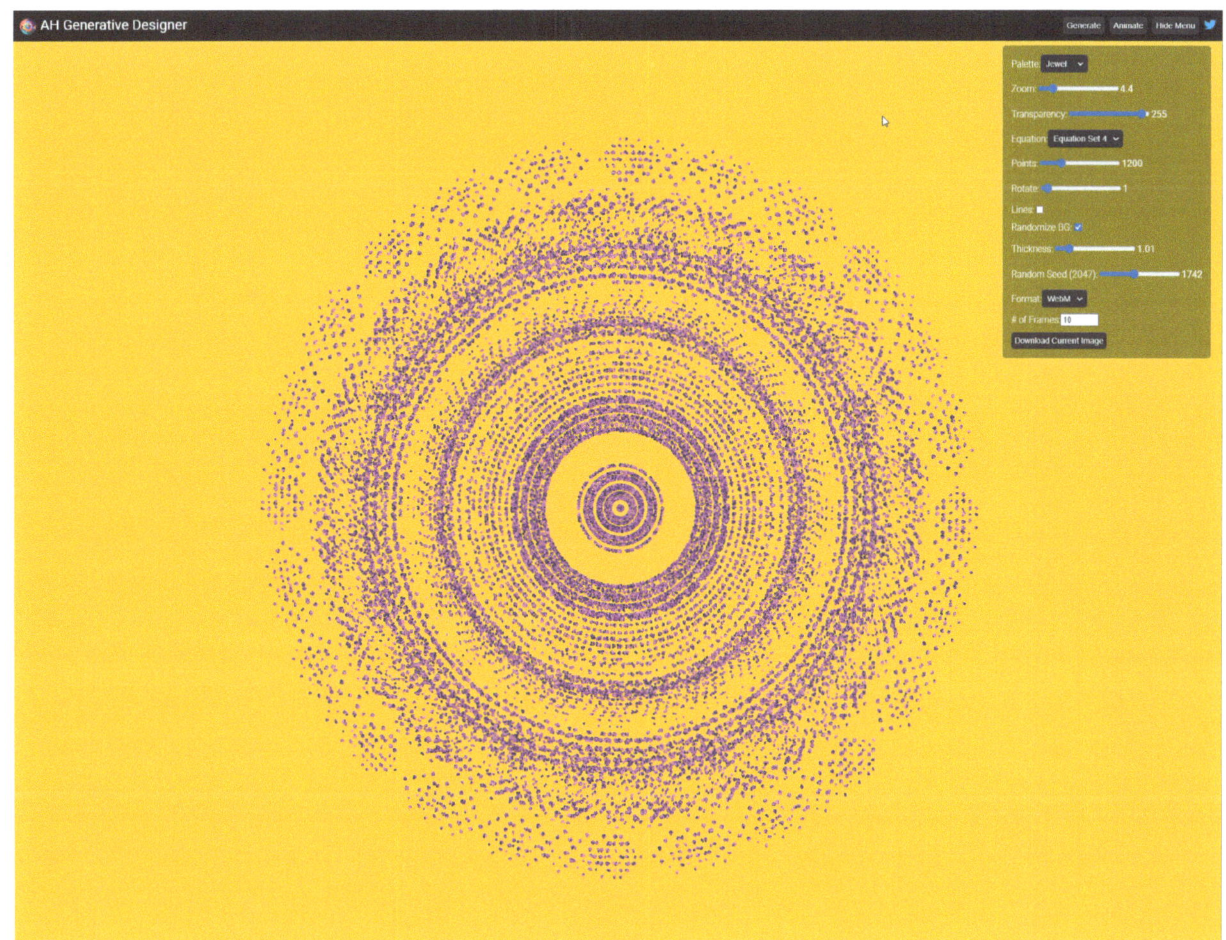

Figure 2.6: A classic Serendipitous Circles rendition with random background colors.

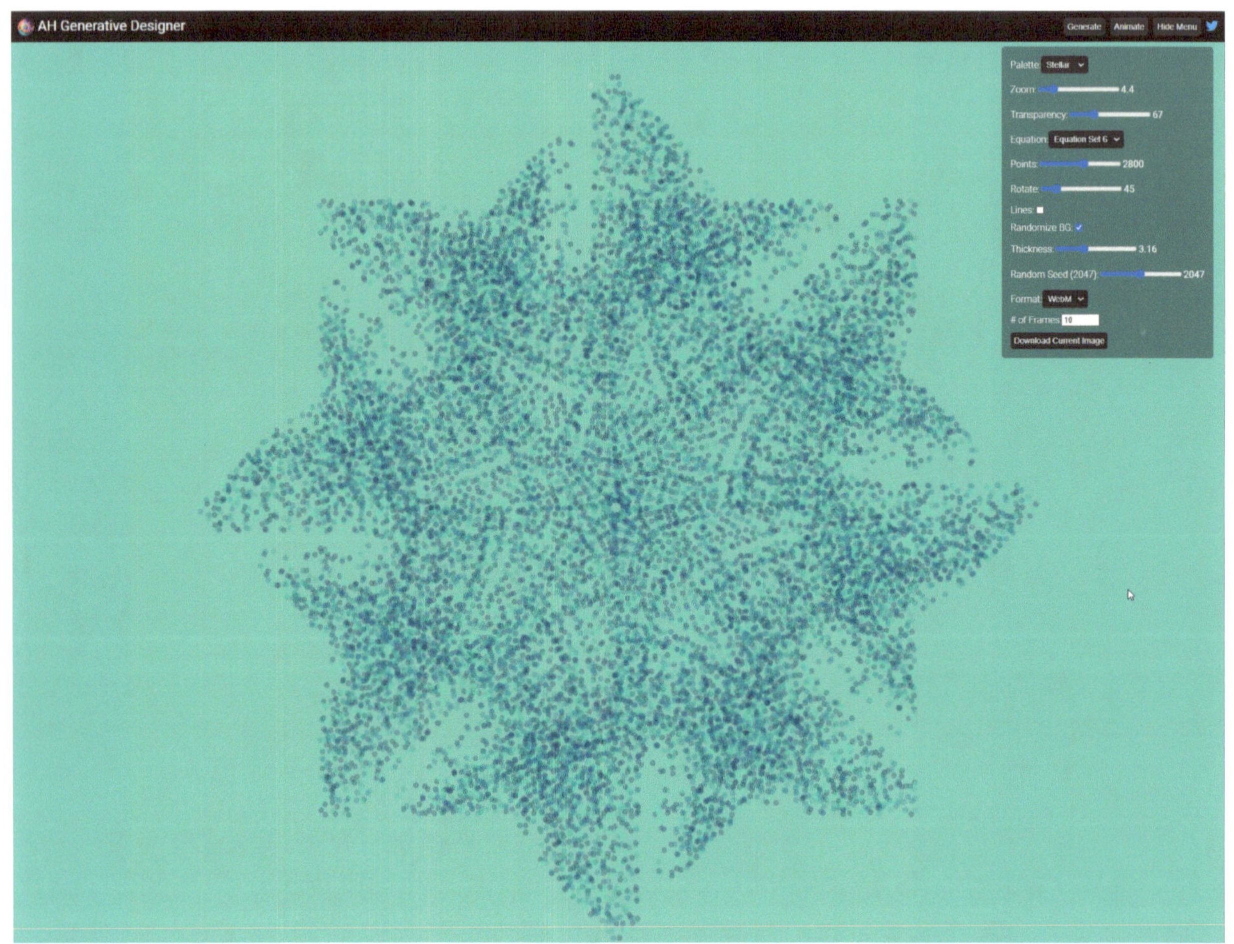

Figure 2.7: There are several palettes to choose from. Here you see the stellar palette used with a fairly thick pen.

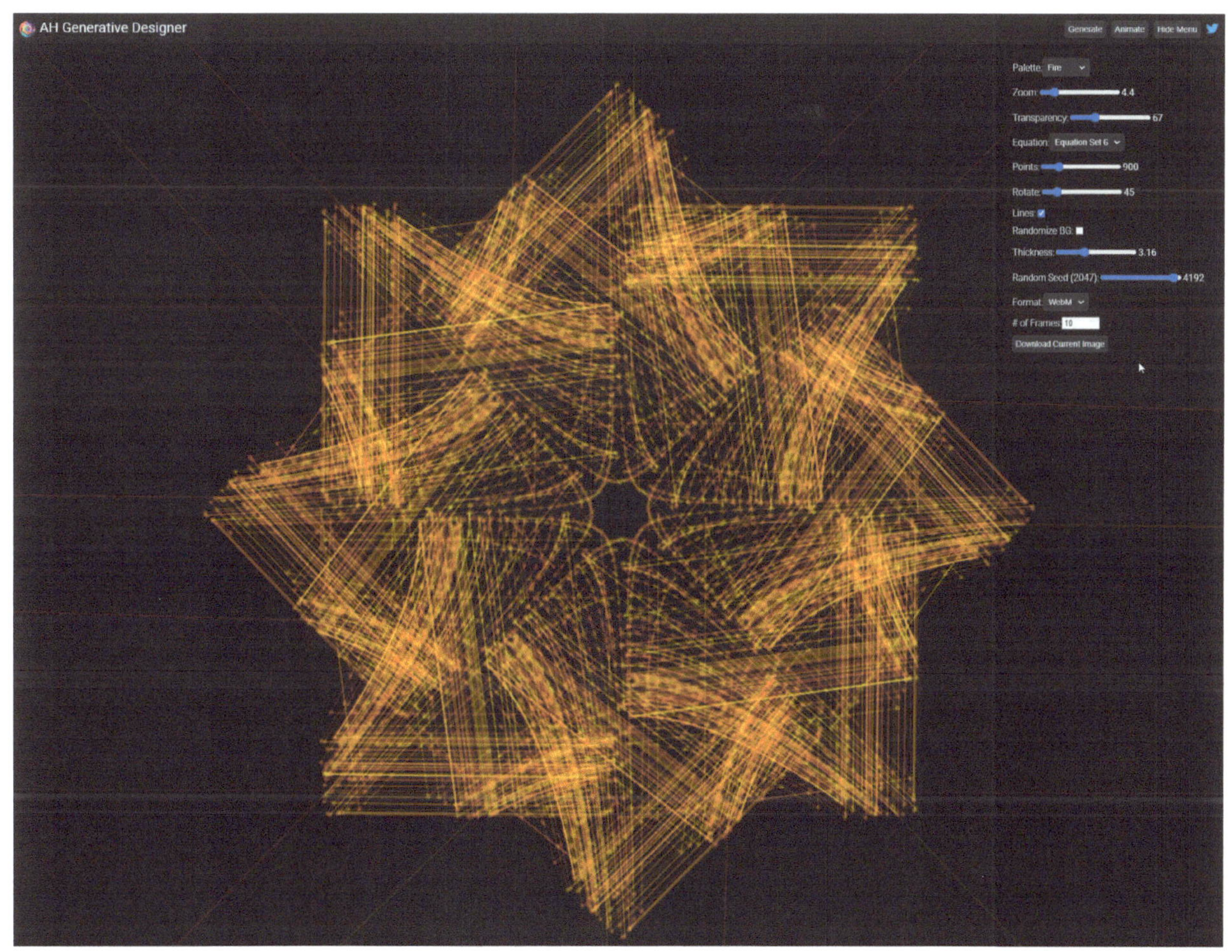

Figure 2.8: The Fire palette on a black background with lines turned on.

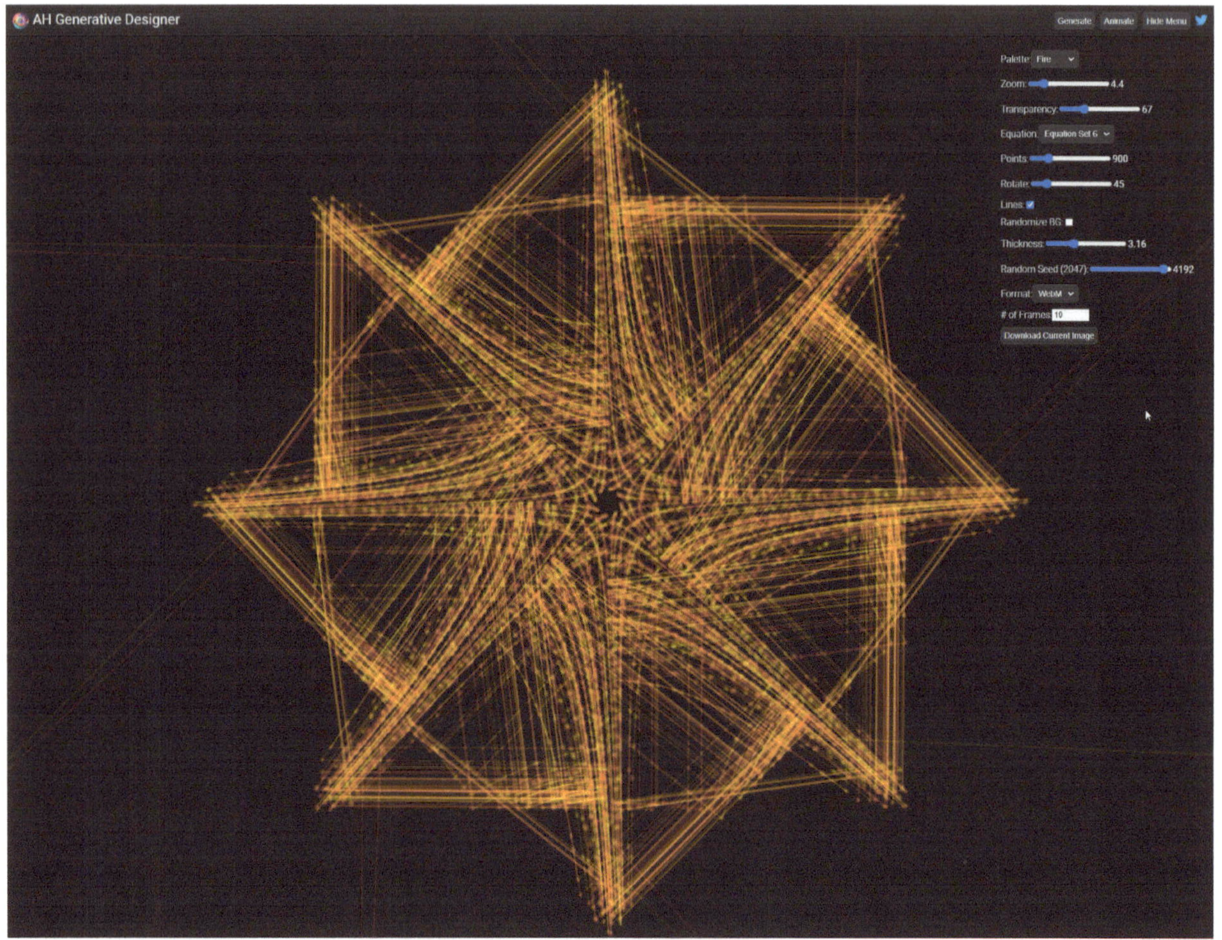

Figure 2.9: All the settings between this image and the last are common. But there is an element of randomness in every drawing. Even with the same settings, you can expect multiple, unique pieces of art!

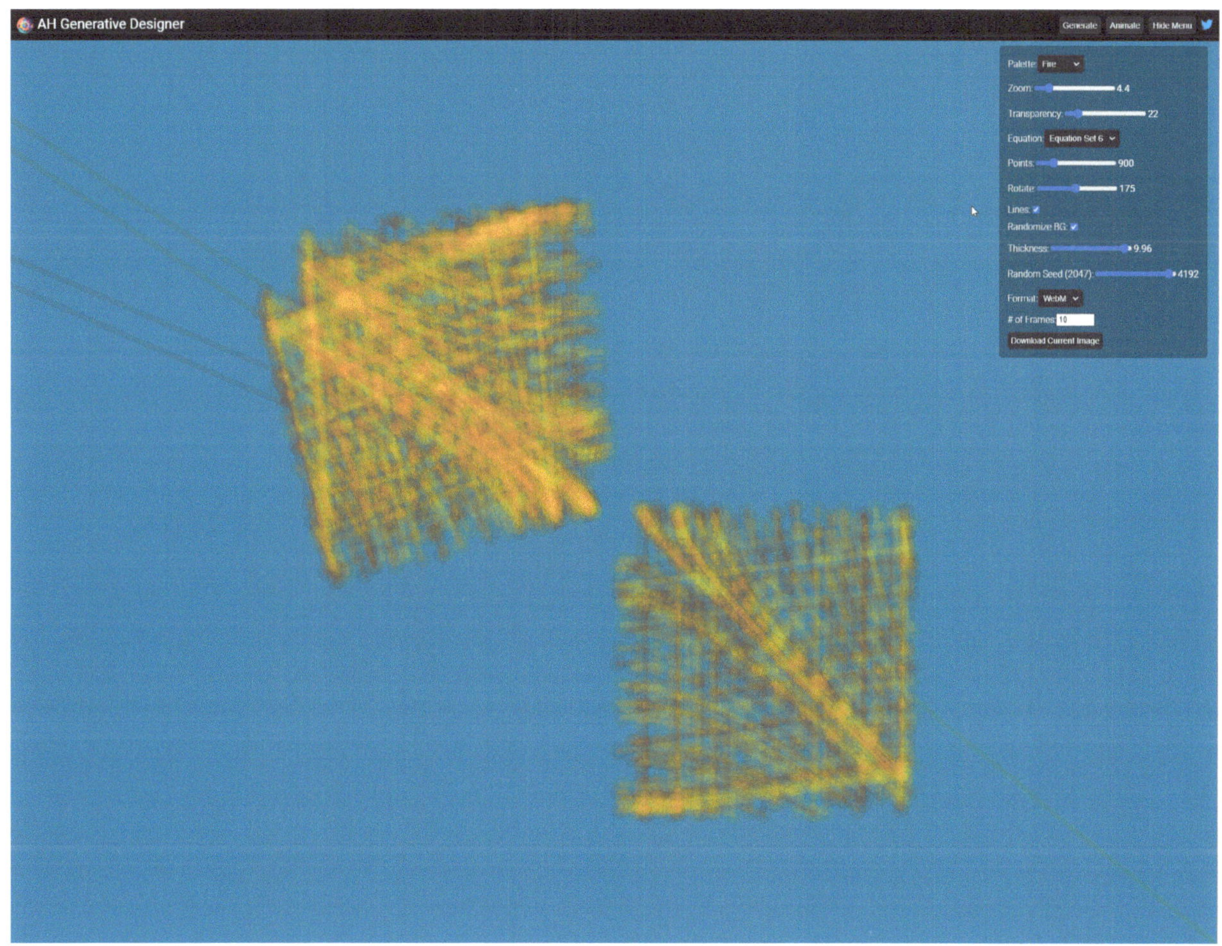

Figure 2.10: Changing to a low transparency and adjusting the rotation angle with random backgrounds can result in some interesting effects. If you continue to click the Generate button there are usually many interesting variations one can inspect in a small amount of time.

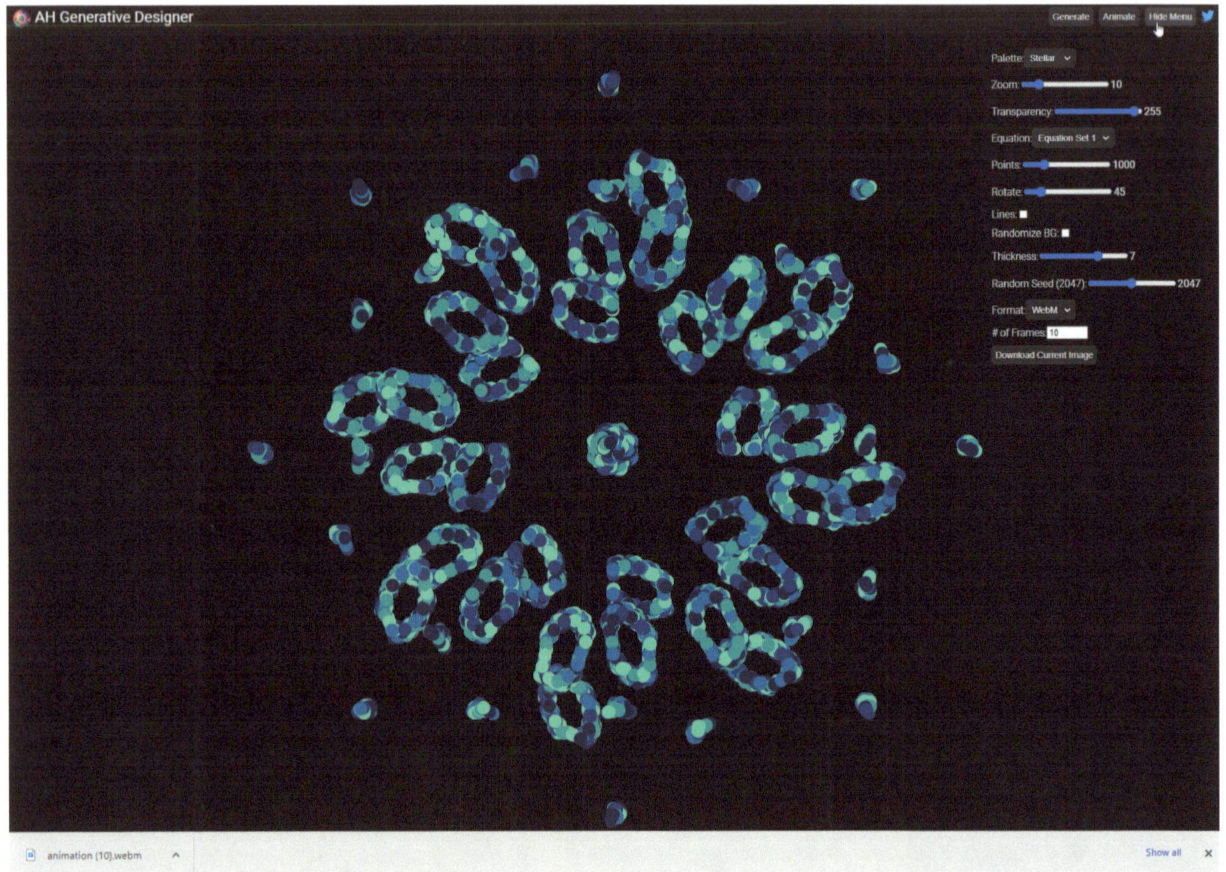

Figure 2.11: You can set animation preferences in the drop-down menu by selecting the number of frames you wish to generate and the type of output you want to create. For example, in this case, a 10 frame animation was produced and it was automatically downloaded to the local system as a .webm Web Media video which is easily playable in standard video playback applications.

Chapter 3

The Generative Designer Implemented in JavaScript

Now that we've done a basic walk-through of the Generative Designer interface, let's take a look under the covers and explore its implementation in JavaScript. In the pages that follow, we will examine key code snippets, review their functionality and purpose so that you can not only understand how this particular Generative Art program has been written, but can also get ideas on how to modify it or create entirely new Generative routines inspired by it.

We will not be covering the HTML and CSS that went into creating the front-end interface, but those are available to you to analyze on your own. An online version of Generative Designer is currently hosted at the following address: `http://openworks.cc/gd`

3.1   Variable Declarations and Setup Function

At the beginning of the code, we define a series of variables and create the canvas.

```
1    let paletteSelect, iterationsSlider, rotationSlider, generateButton;
2    let windowWidth = window.innerWidth;
3    let windowHeight = window.innerHeight;
```

The variables `paletteSelect`, `iterationsSlider`, `rotationSlider`, and `generateButton` will be used to store references to the user interface elements.

```javascript
function setup() {
    createCanvas(windowWidth, windowHeight);
    background(0);

    // Get the UI elements
    paletteSelect = document.getElementById('palette');
    iterationsSlider = document.getElementById('iterations');
    rotationSlider = document.getElementById('rotation');
    generateButton = document.getElementById('generate');

    // Add event listeners to UI elements
    iterationsSlider.addEventListener('input', (e) => document.
        getElementById('iterationsValue').innerText = e.target.value);
    rotationSlider.addEventListener('input', (e) => document.
        getElementById('rotationValue').innerText = e.target.value);
    generateButton.addEventListener('click', generateArt);
    document.getElementById('lineWidth').addEventListener('input', (
        event) => {
        document.getElementById('lineWidthValue').innerText = event.
            target.value;
    });

    document.getElementById('increment').addEventListener('input', (
        event) => {
        document.getElementById('incrementValue').innerText = event.
            target.value;
    });
}
```

The `setup()` function initializes the HTML5 canvas upon which the drawing will be created and sets up event listeners for the user interface elements. When users interact with the UI elements within the browser, the appropriate listeners will trigger and update values so that they can be used as inputs within the Generative Designer program.

3.2 Window Resized Function

```
1    function windowResized() {
2            resizeCanvas(windowWidth, windowHeight);
3        }
```

The `windowResized()` function ensures that the canvas is resized whenever the browser window size changes. Users can resize the browser on laptops and desktops, or rotate the browser on a mobile device. This code accounts for those changes in dimensions.

3.3 Draw and RunFrames Functions

```
1    function draw() {}
2
3    async function runFrames() {
4    // ...
5        }
```

These two functions can currently be thought of as stub functions that will be completed later. The empty `draw()` function is included as a placeholder, while the `runFrames()` function generates and plays back a series of images in support of the future animation functionality. We won't get into the details of these yet as work is incomplete.

3.4 GenerateArt Function

```
1    function generateArt() {
2            // Get the selected palette and other settings
3            const selectedPalette = paletteSelect.value;
4            const lineWidth = parseFloat(document.getElementById('lineWidth').
                ↪ value);
5            const increment = parseInt(document.getElementById('increment').
                ↪ value);
6            const connectLines = document.getElementById('connectLines').checked
                ↪ ;
```

```javascript
7    const randomBackground = document.getElementById('randomBackground')
         ↪ .checked;
8    const iterations = int(iterationsSlider.value);
9    const rotationDegrees = int(rotationSlider.value);
```

The `generateArt()` function begins by retrieving the user-selected settings from the UI elements. Think of this as the "glue" code that connects the user interface to the drawing engine.

```javascript
1    // Define palettes and background colors
2    const palette = {
3            // ...
4    }[selectedPalette];
5
6    const bgs = [
7    color(0, 0, 0),
8    color(255, 204, 0),
9    color(255, 255, 255),
10   color(random(255), random(255), random(255))
11   ];
```

Next, we define the color palettes and background colors. The selected palette is retrieved from the **palette** object based on the user's choice. There are a number of palettes already configured in Generative Designer and the multiple colors that are part of each palette have been chosen so that they work well together. Additional palettes can easily be added or loaded from an external source. Some great ideas on palettes that work well together can be found here: `http://colormind.io/`

```javascript
1    // Set the background color
2    if (randomBackground) {
3            background(random(bgs));
4    } else {
5            background(0);
6    }
7    smooth(4);
```

The background color is set based on the user's choice of random or black background. The **smooth()** function is called to improve the appearance of lines and shapes. The list of background colors is limited and can be extended easily by you.

```
const rotf_adder = rotationDegrees;
const pal = palette;
let c = random(pal);
let x = random(65536/4);
let y = random(65536/4);
const origx = x;
const origy = y;
let oldx = x;
let oldy = y;
const andval = increment;
let rotf = 0;

translate(width / 2, height / 2);
```

The variables for rotation, color, and position are initialized, and the canvas is translated —or shifted—so that the origin is at the center of the canvas.

```
while (rotf < 360) {
    x = origx;
    y = origy;
    for (let i = 0; i < iterations; i++) {
        c = random(pal);
        const a = random(65, 255);
        c.setAlpha(a);
        strokeWeight(lineWidth * random(0.4, 2));
        stroke(c);
        line(x/6, y/6, x/6, y/6);
        //c.setAlpha(255);
        stroke(c);
        if (connectLines) {
            line(oldx/6, oldy/6, x/6, y/6);
        }

        oldx = x;
        oldy = y;
```

```
19
20                            x = x - y & andval;
21                            y = y + (x / 2) & andval;
22                    }
23
24                 rotf = rotf + rotf_adder;
25                 rotate(radians(rotf));
26            }
```

The main loop iterates over the angles (from 0 to 360) and calculates the position of each point that is to be plotted using the generative algorithm. A series of lines is drawn based on the user-selected settings received from the user interface. The canvas is rotated at each step to create the final pattern. This rotation is where the images derive their symmetry about the center.

You can modify the program in many ways to create different results. For example, translate the canvas so that the center about which points are plotted is not the visual center of the image. Similarly, where the rotation increments (rotf_adder) is randomly chosen from a range for each iteration.

I hope in reviewing this chapter, you have gained an appreciation for the JavaScript implementation of the Generative Designer. The code is so small and simple that it almost invites endless modification and creative play!

Chapter 4

Generative Art Gallery

Generative Designer can be used to create a wide range of art pieces that are each unique, both in color and style, and also in subtle details. Two pieces may initially look alike, but when you look at the details you will inevitably find that there is a range of differences. Each piece of art comes with its own unique signature... like snowflakes!

This chapter is a visual tour of a very small range of variations that are possible to achieve with Generative Designer. They do not define the boundaries of the variations and artistic outcomes you can achieve, but are a small sampling of what can be done.

Let us walk through the gallery.

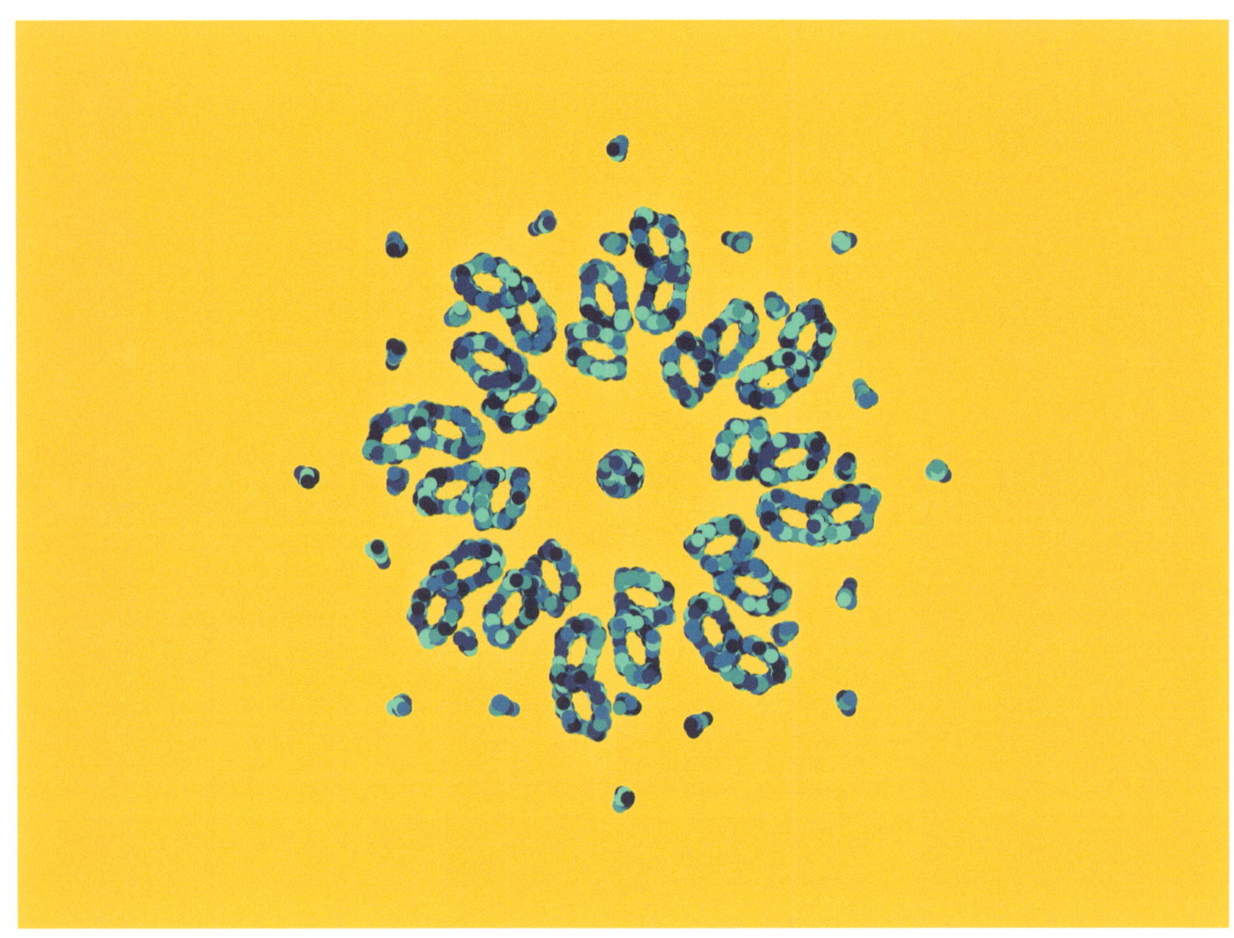

Figure 4.1: Sueños (Dreams)

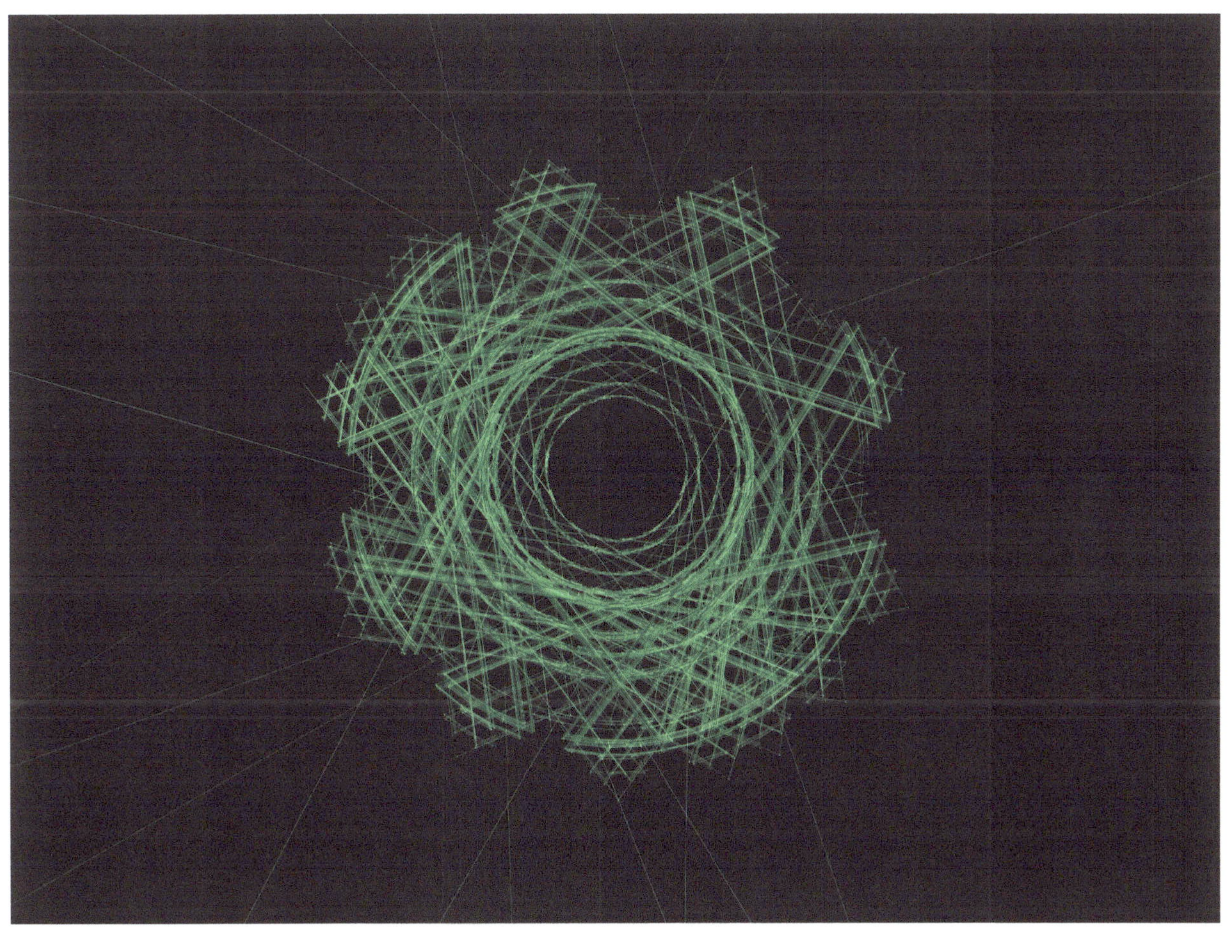

Figure 4.2: Einsamkeit (Loneliness)

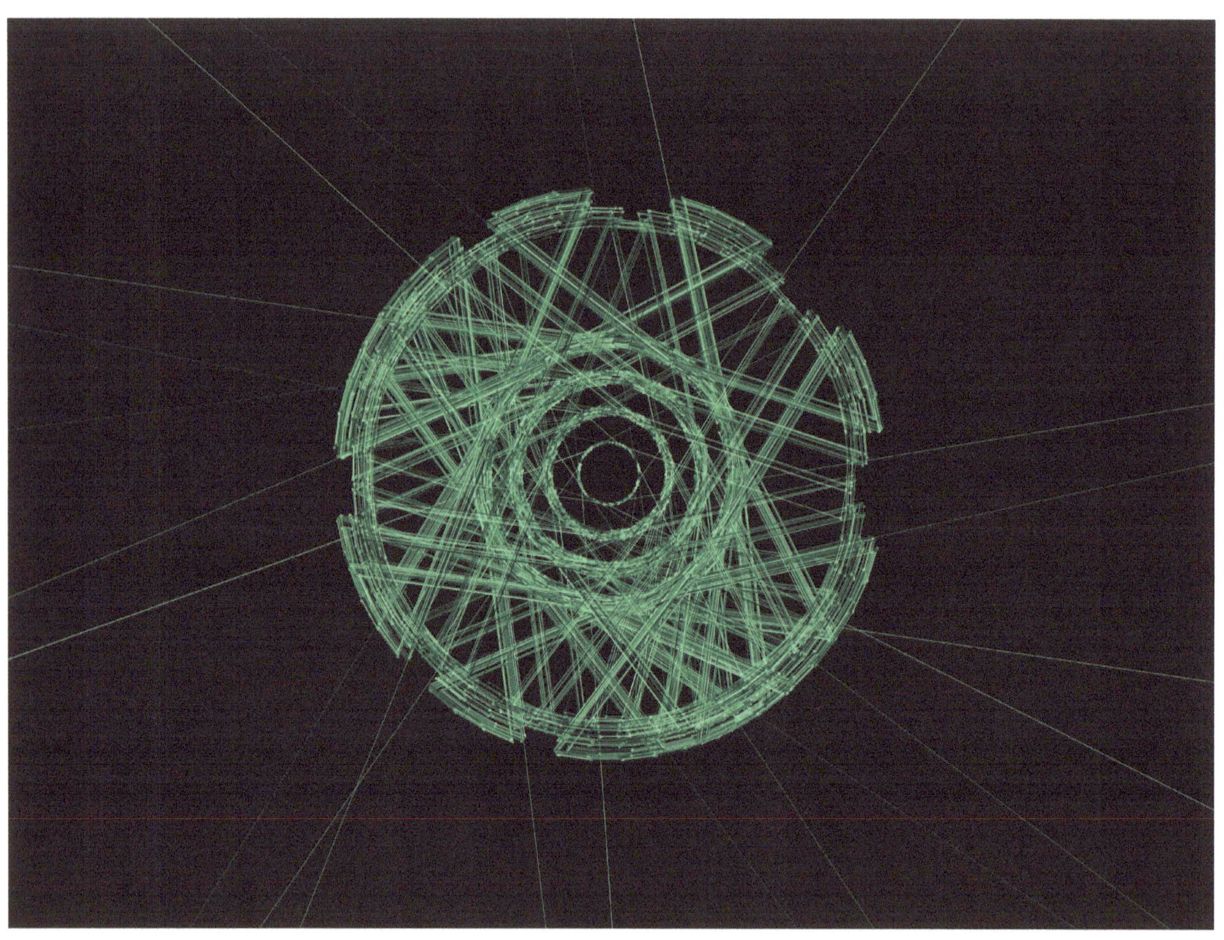

Figure 4.3: Serendipia (Serendipity)

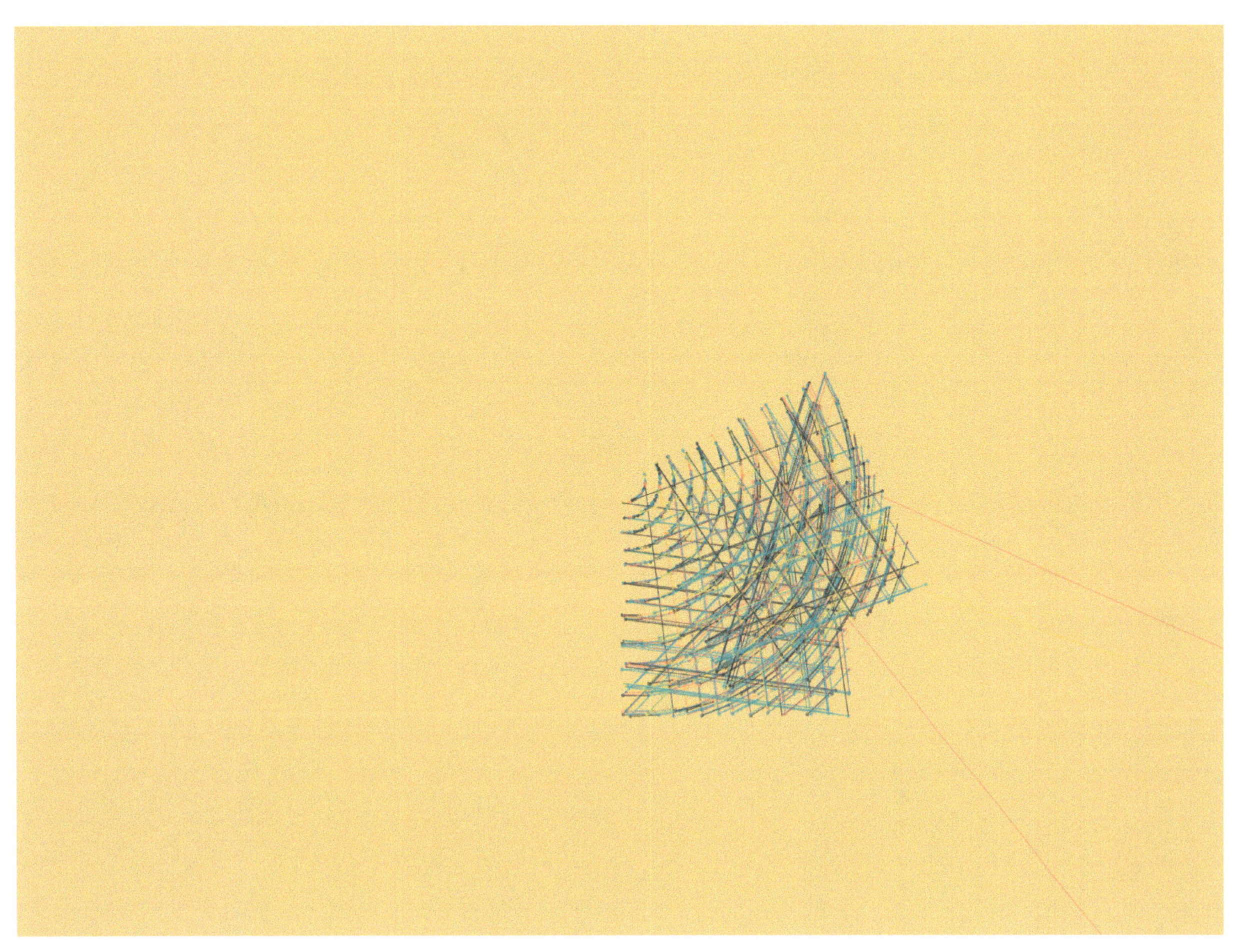

Figure 4.4: وردة (Rose)

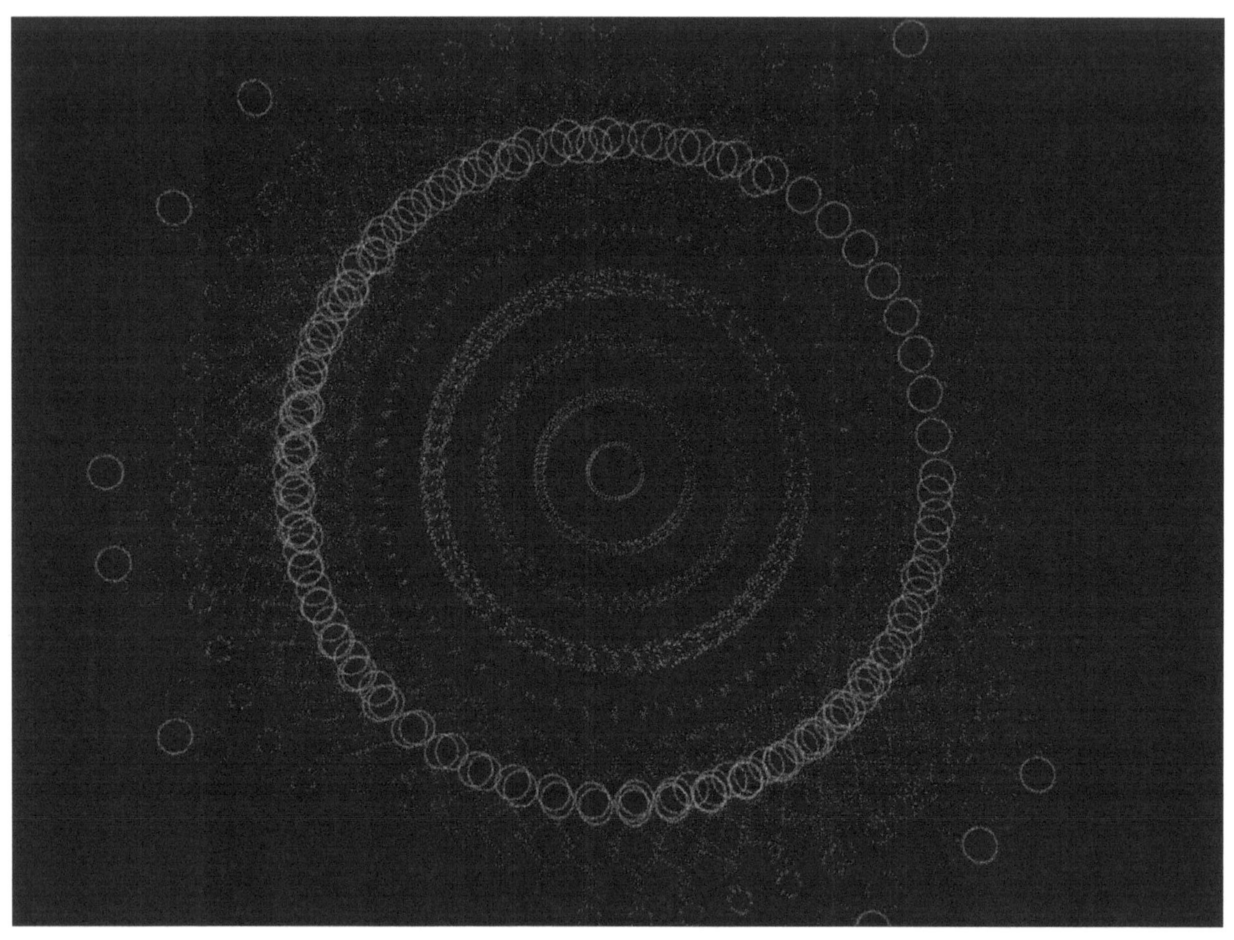

Figure 4.5: Wanderlust (Wanderlust)

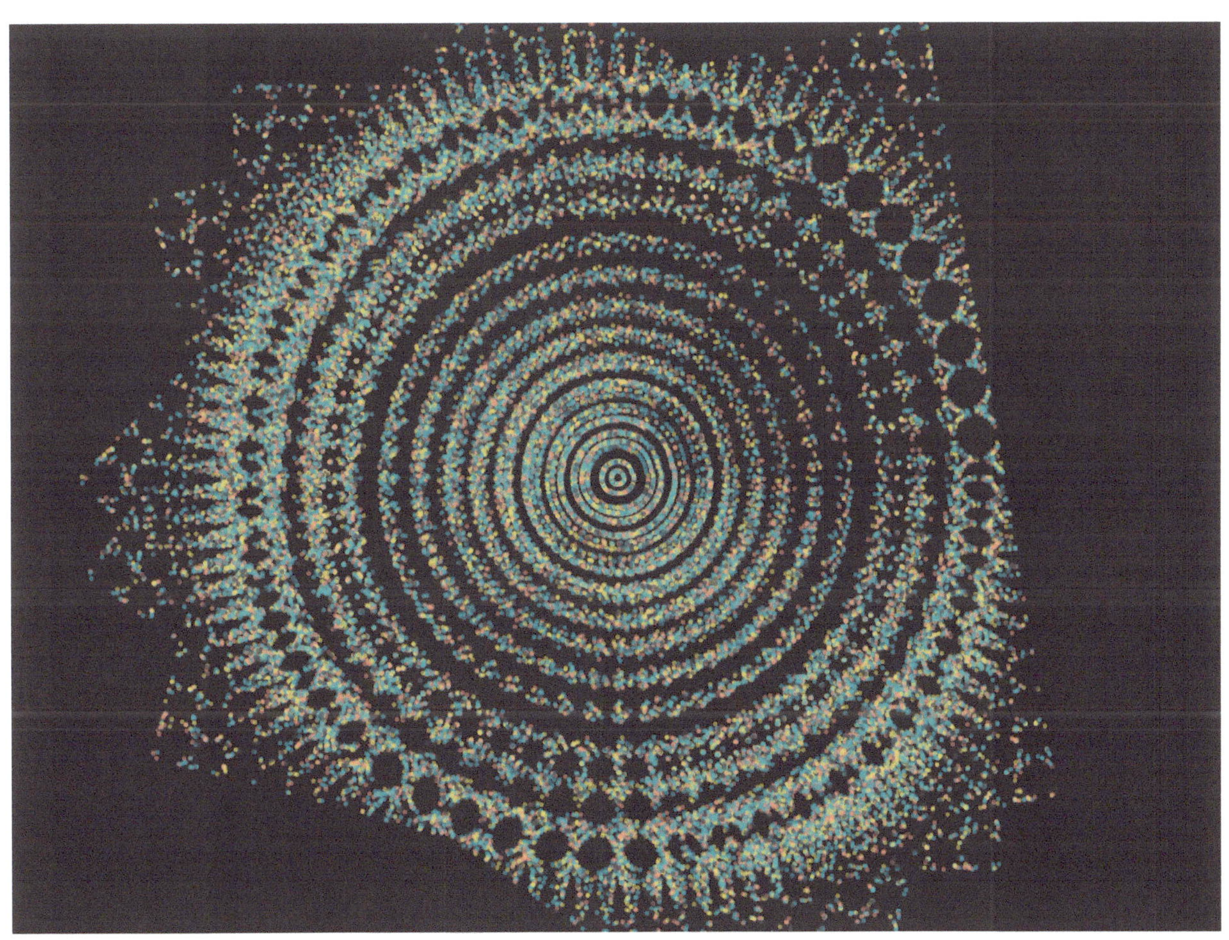

Figure 4.6: Flüstern (Whisper)

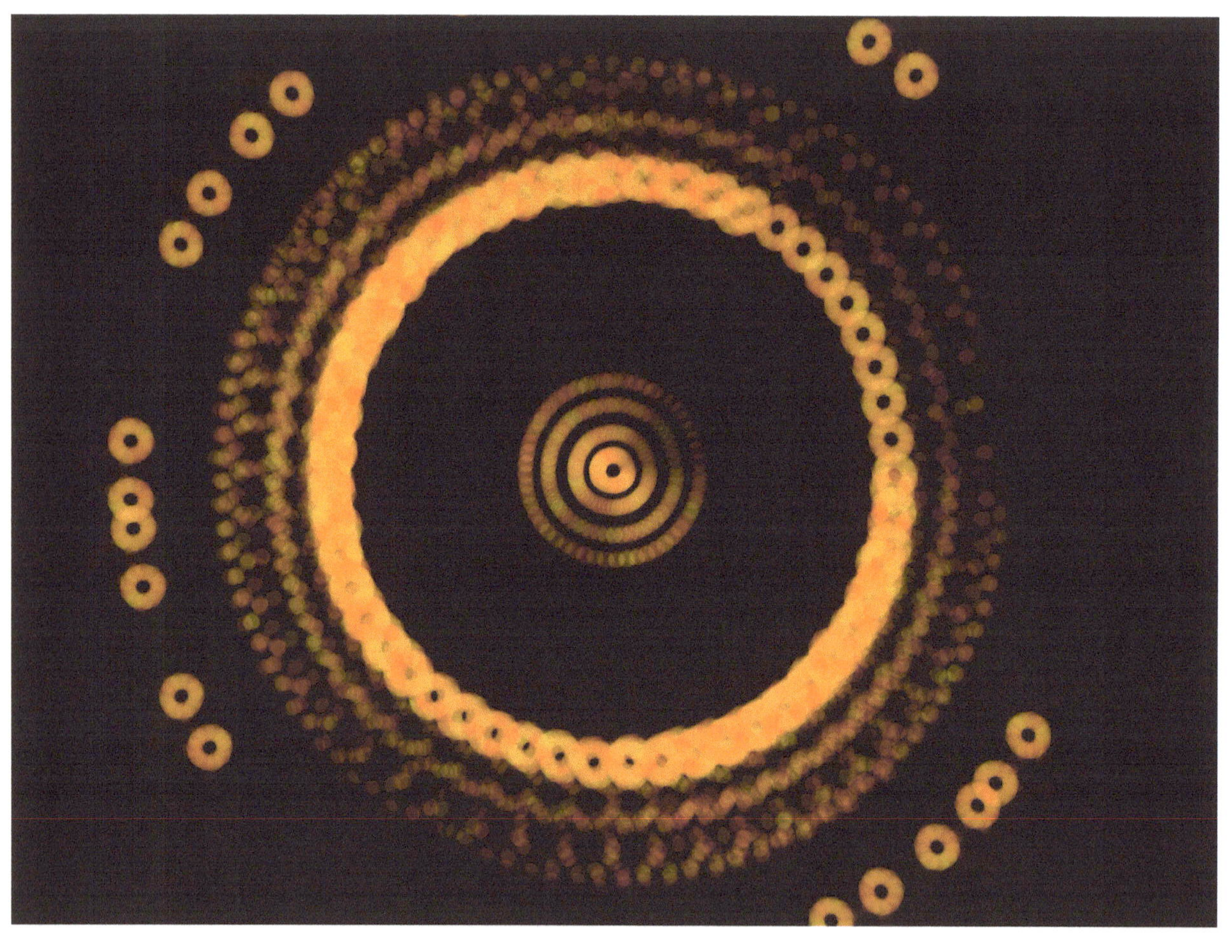

Figure 4.7: ألوان (Colors)

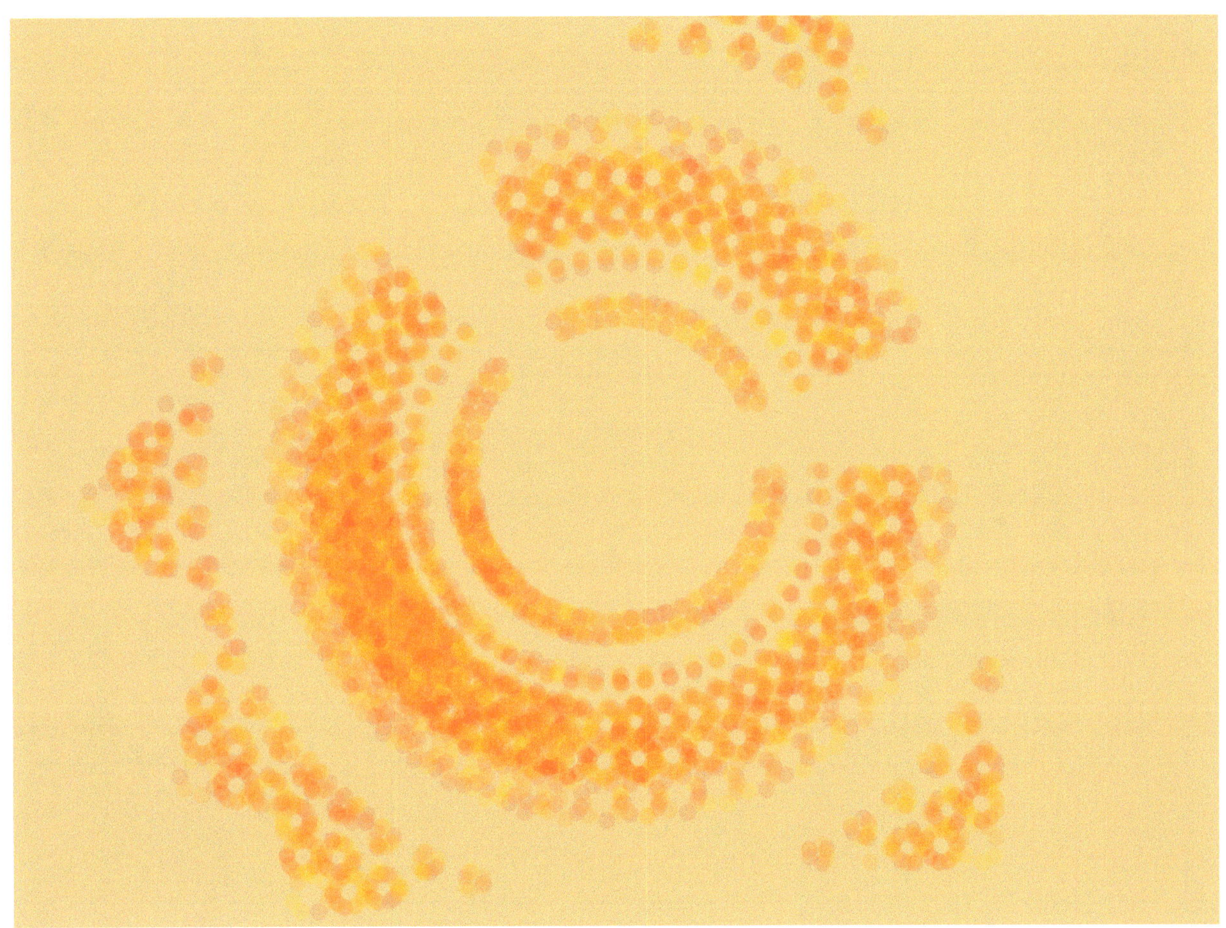

Figure 4.8: Esperanza (Hope)

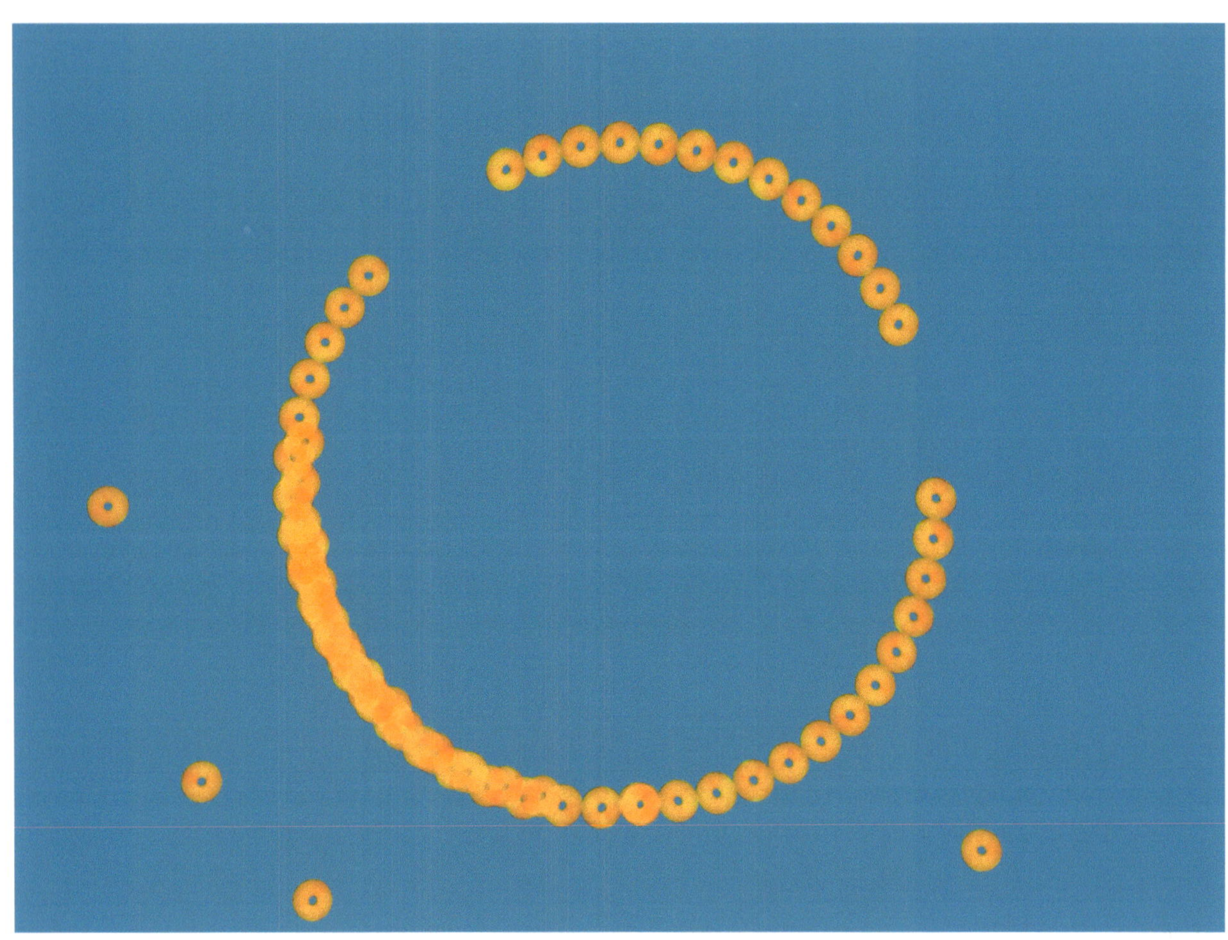

Figure 4.9: Sternenlicht (Starlight)

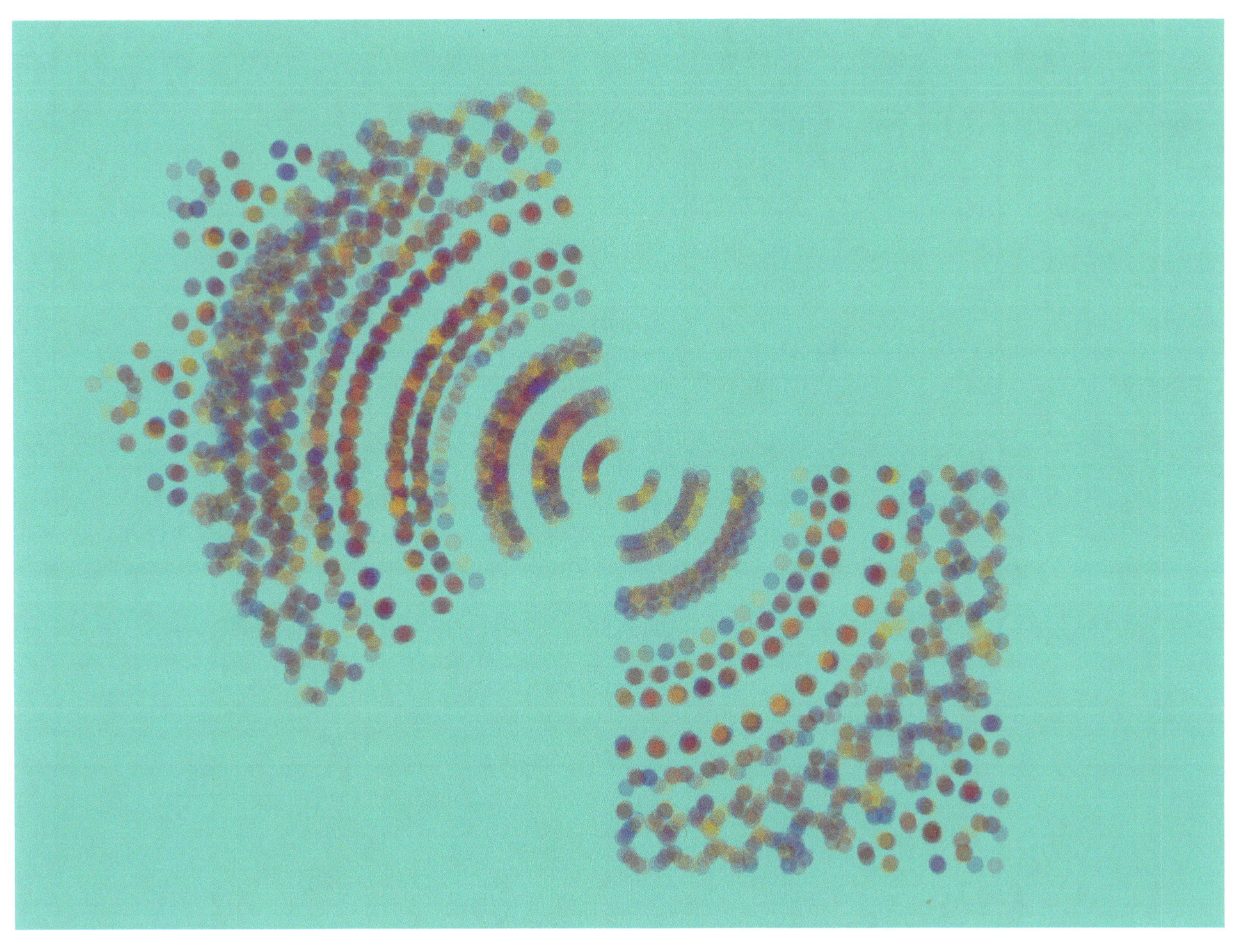

Figure 4.10: إشراقة (Glimmer)

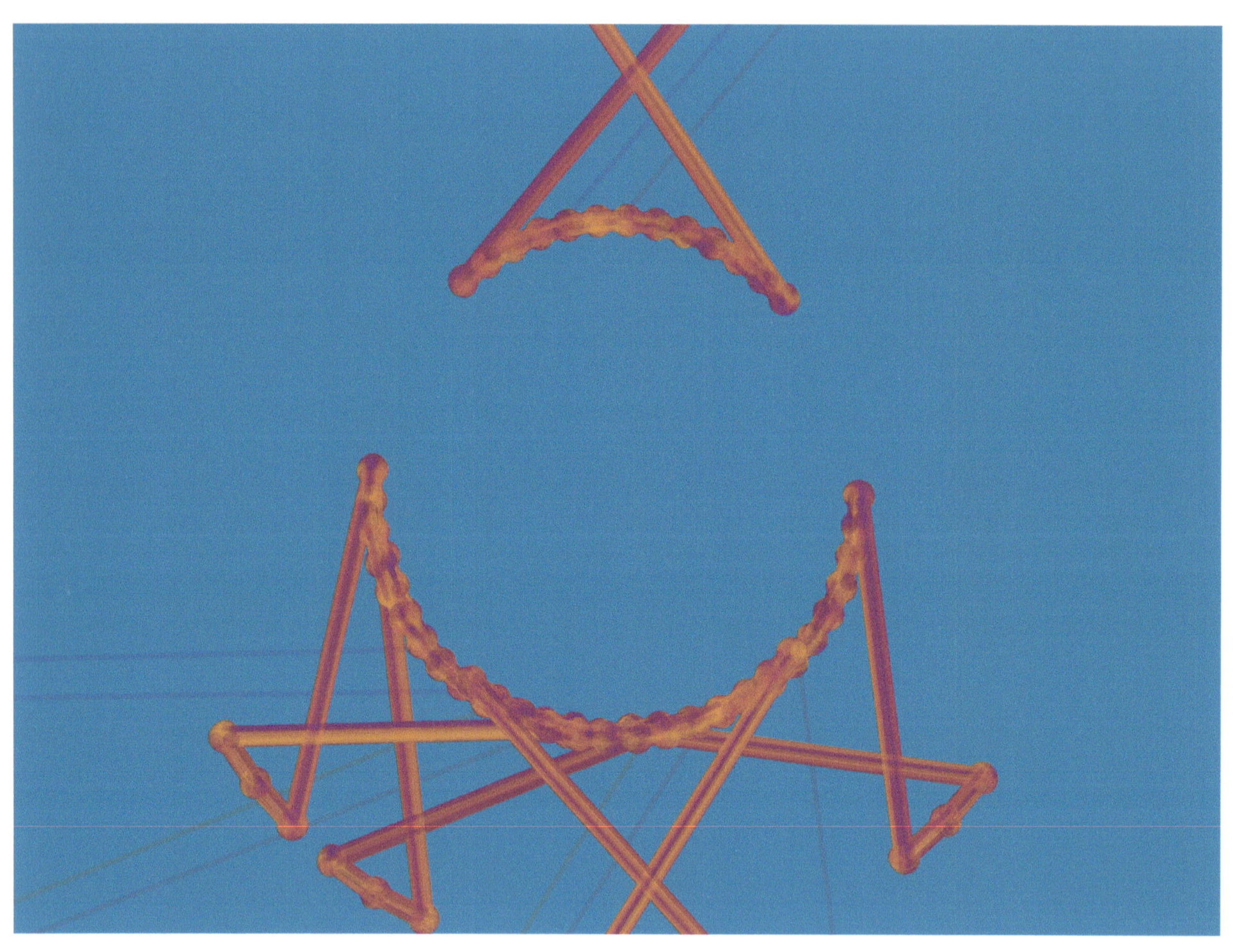

Figure 4.11: Armonía (Harmony)

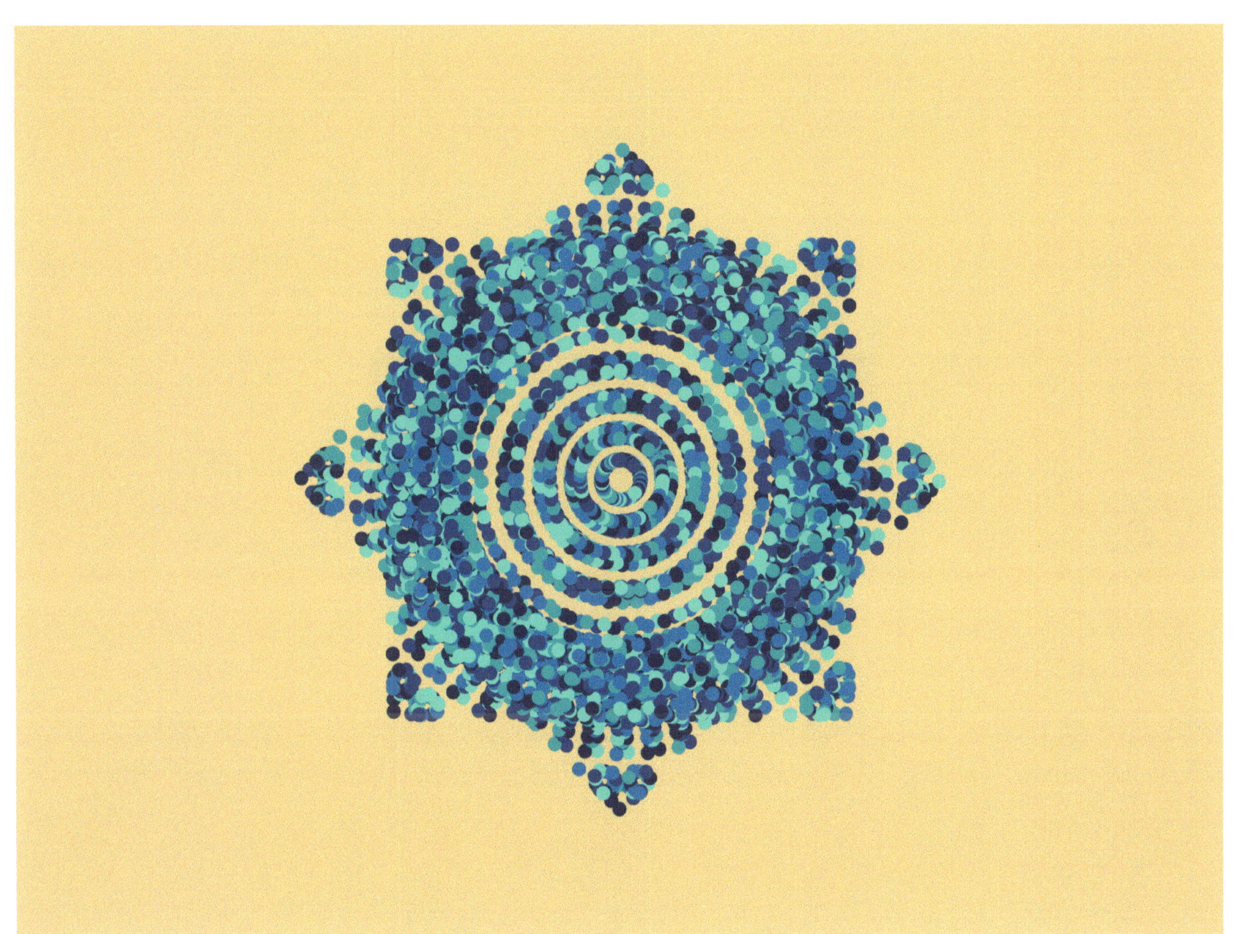

Figure 4.12: Frieden (Peace)

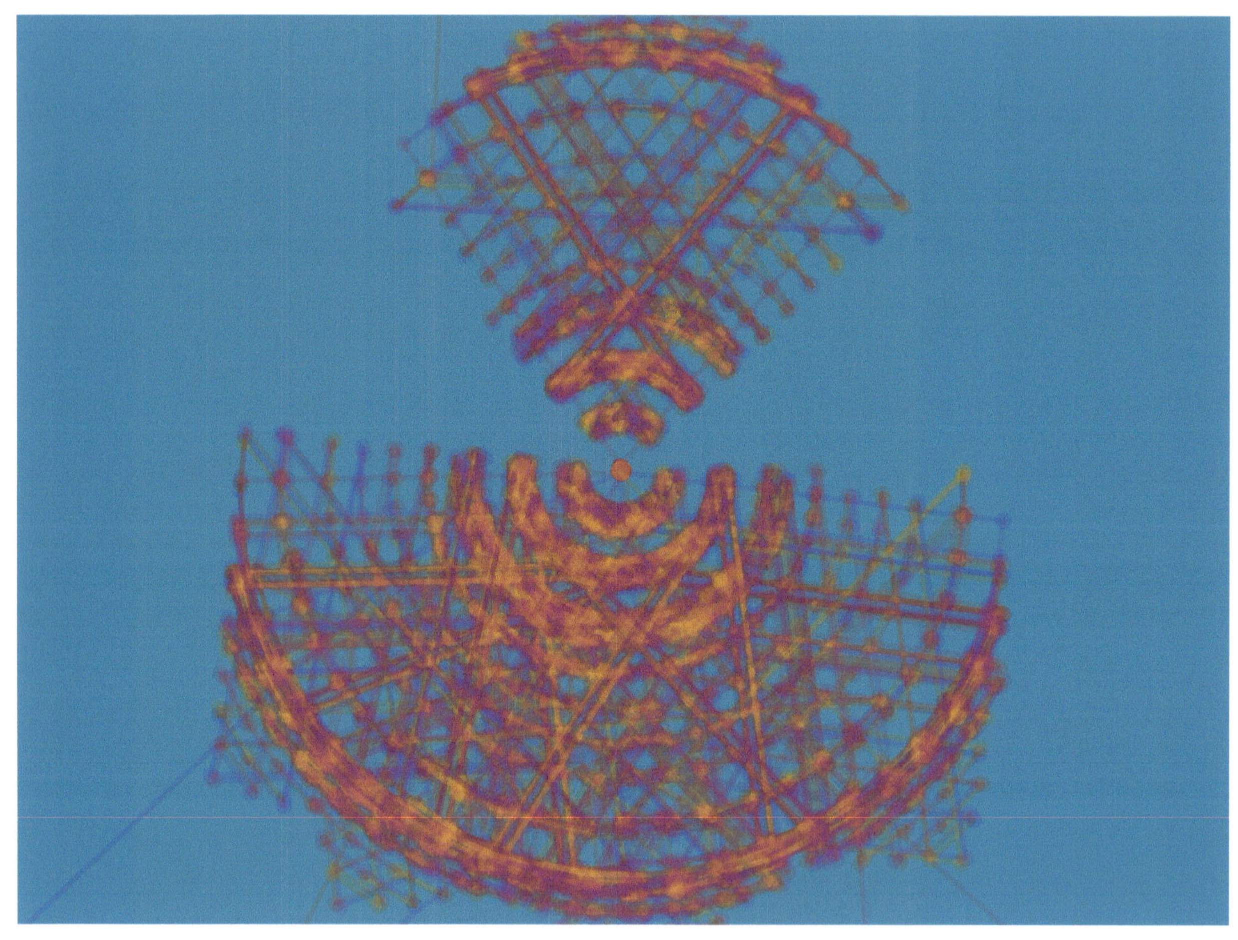

Figure 4.13: Reflektion (Reflection)

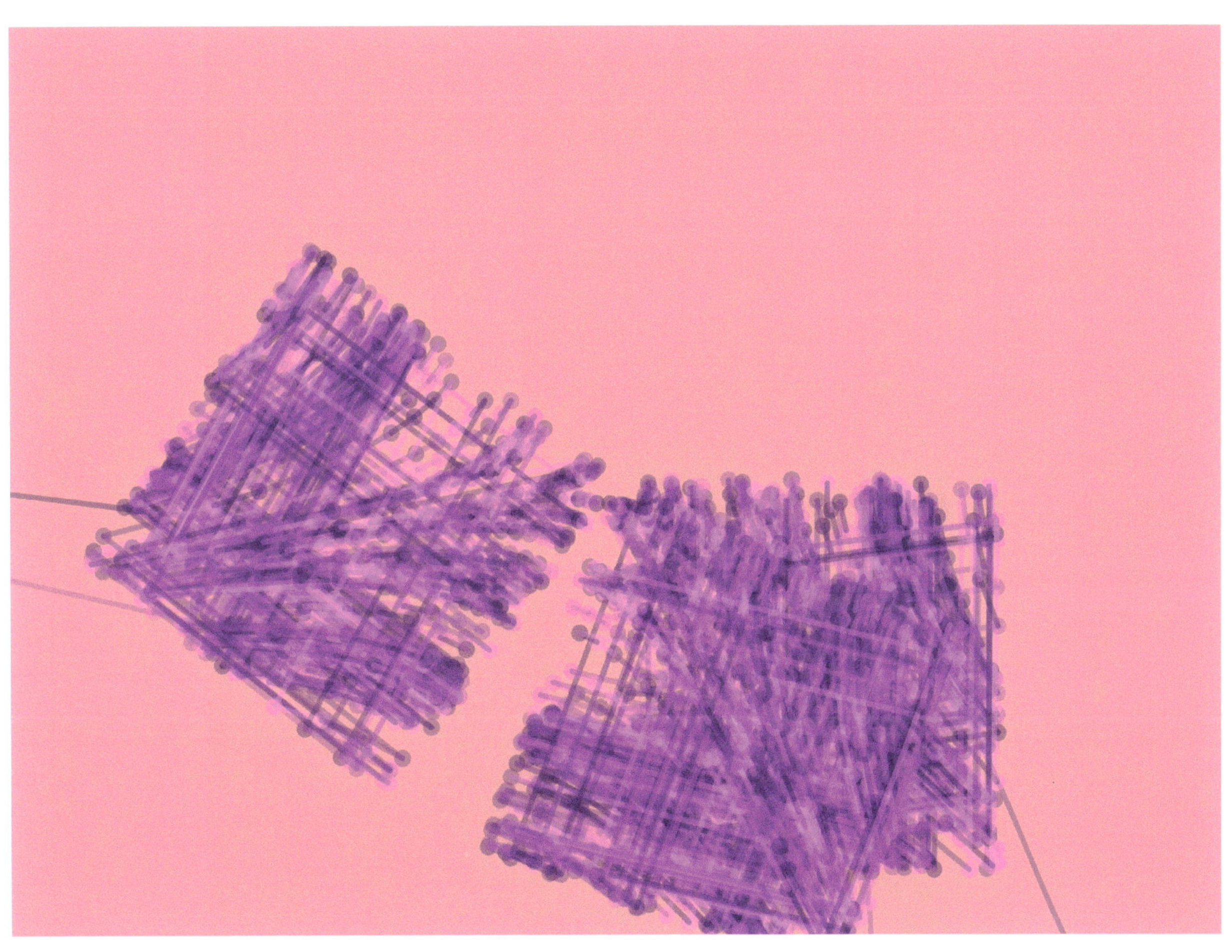

Figure 4.14: نسيم (Breeze)

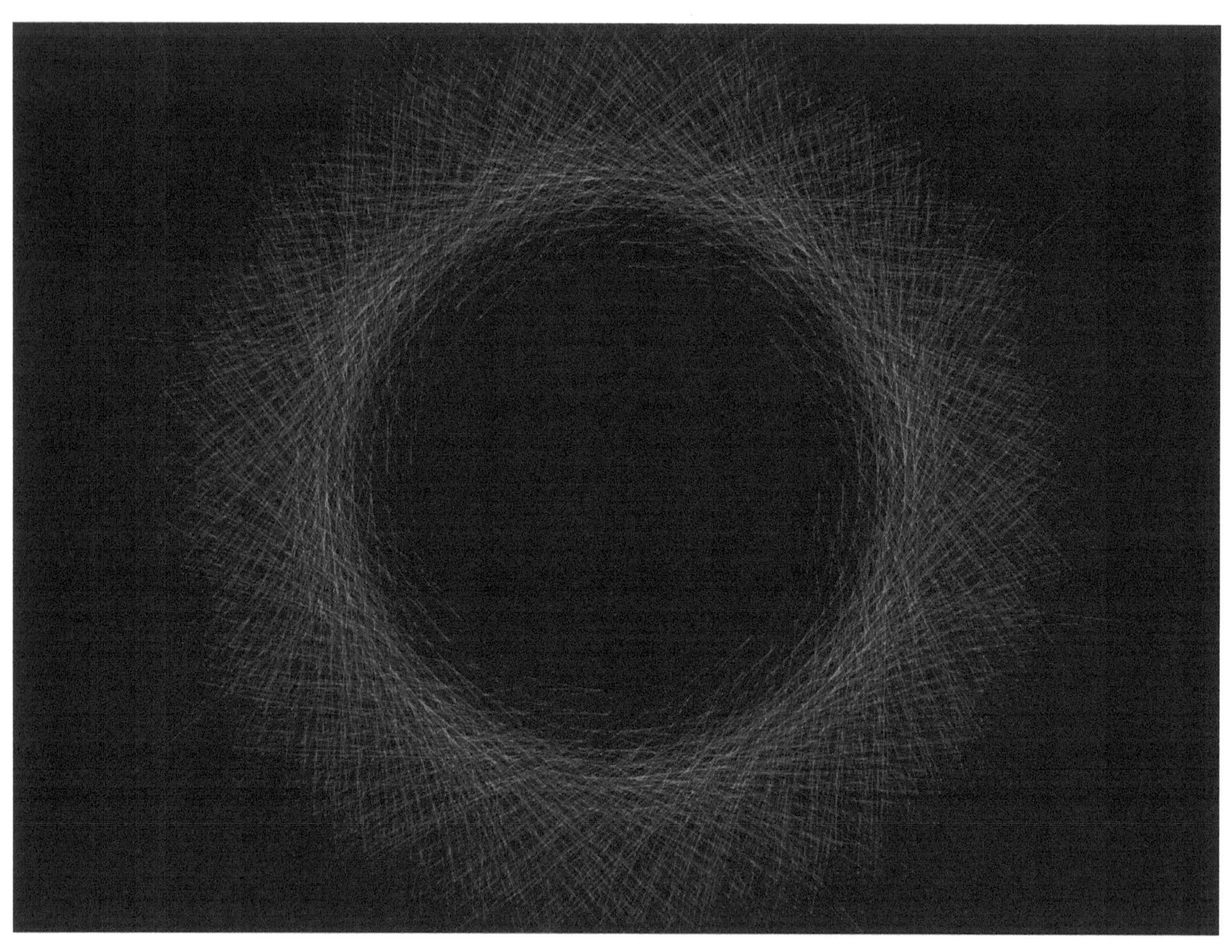

Figure 4.15: Rêverie (Daydream)

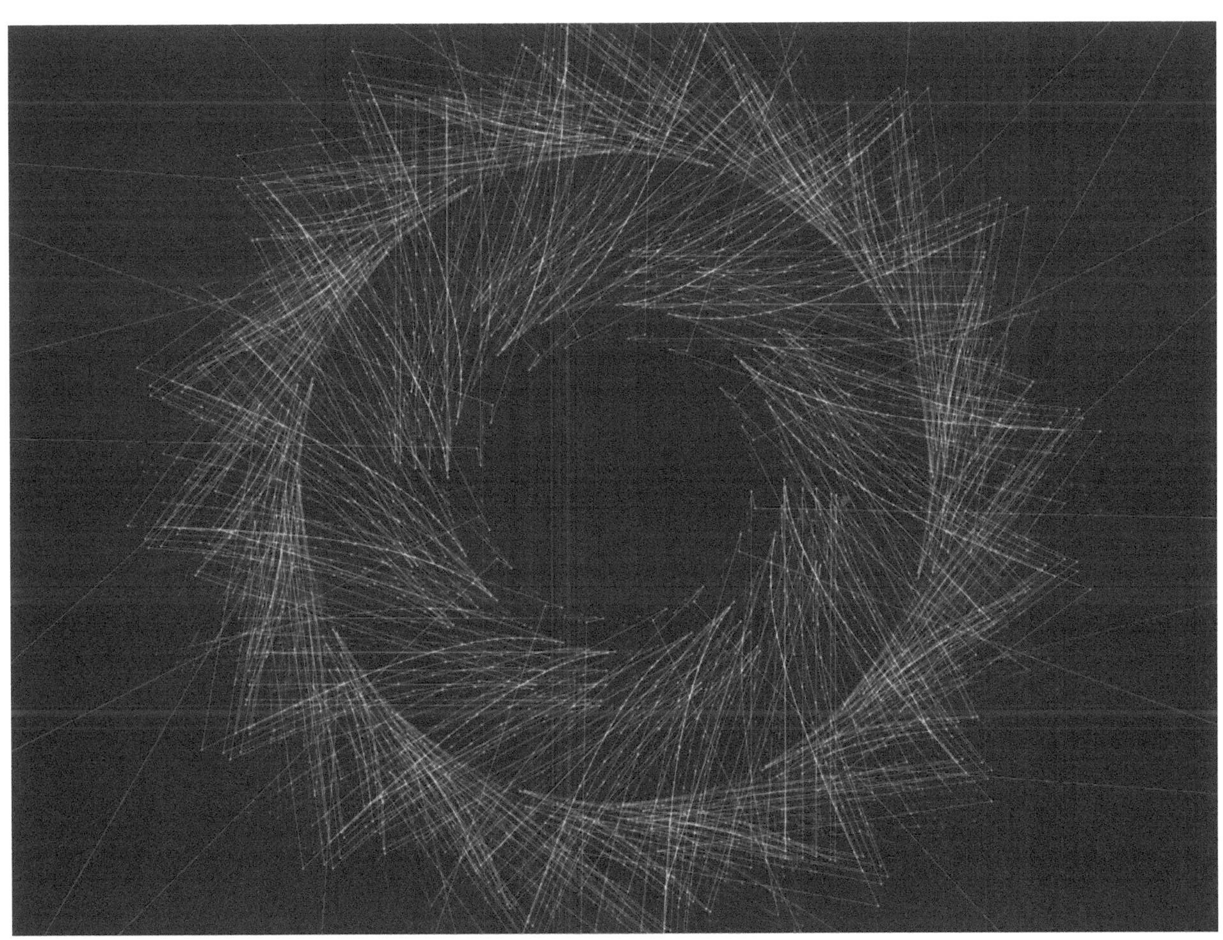

Figure 4.16: Atardecer (Sunset)

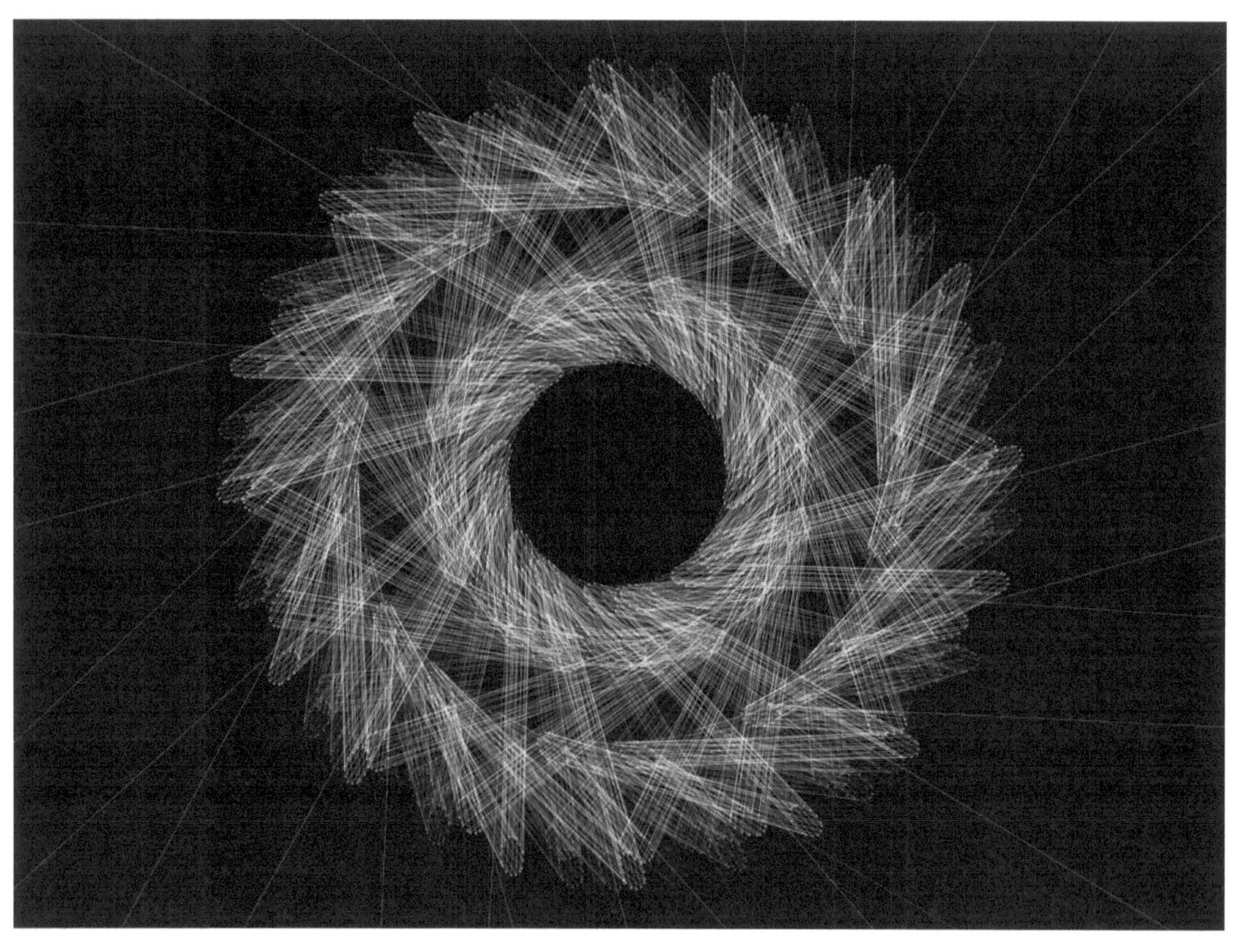

Figure 4.17: Abstraktion (Abstraction)

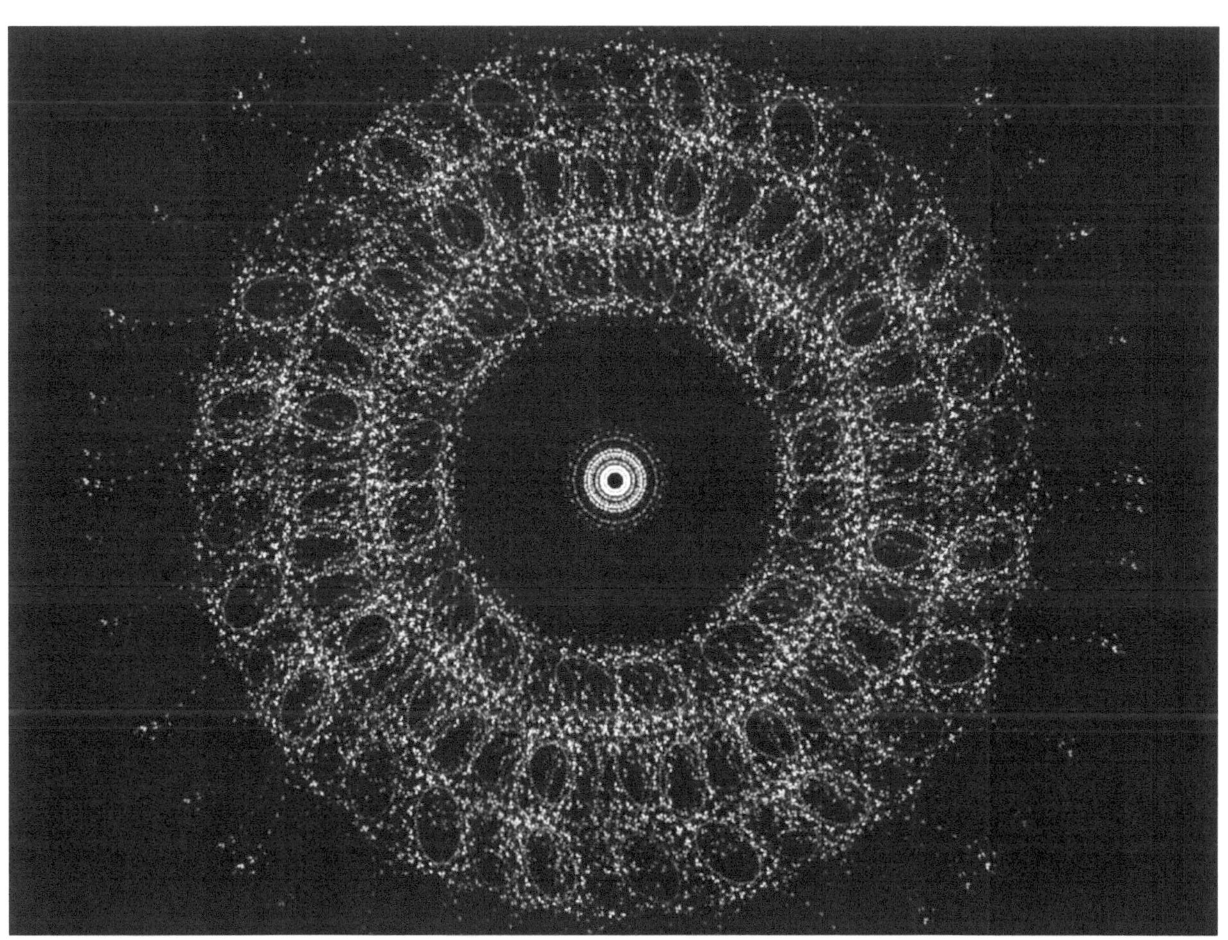

Figure 4.18: زهور (Flowers)

Figure 4.19: Verwandlung (Transformation)

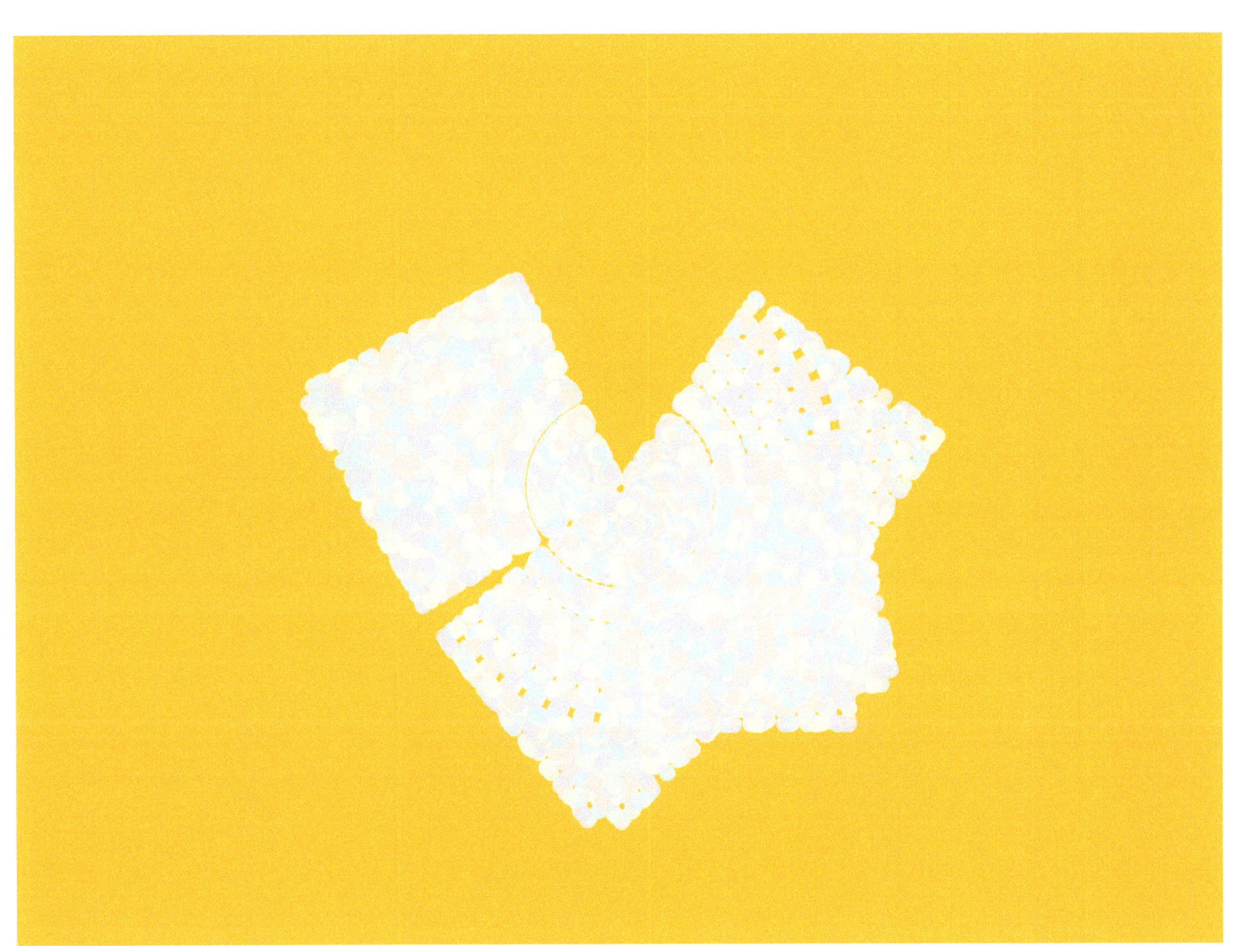

Figure 4.20: Enigma (Enigma)

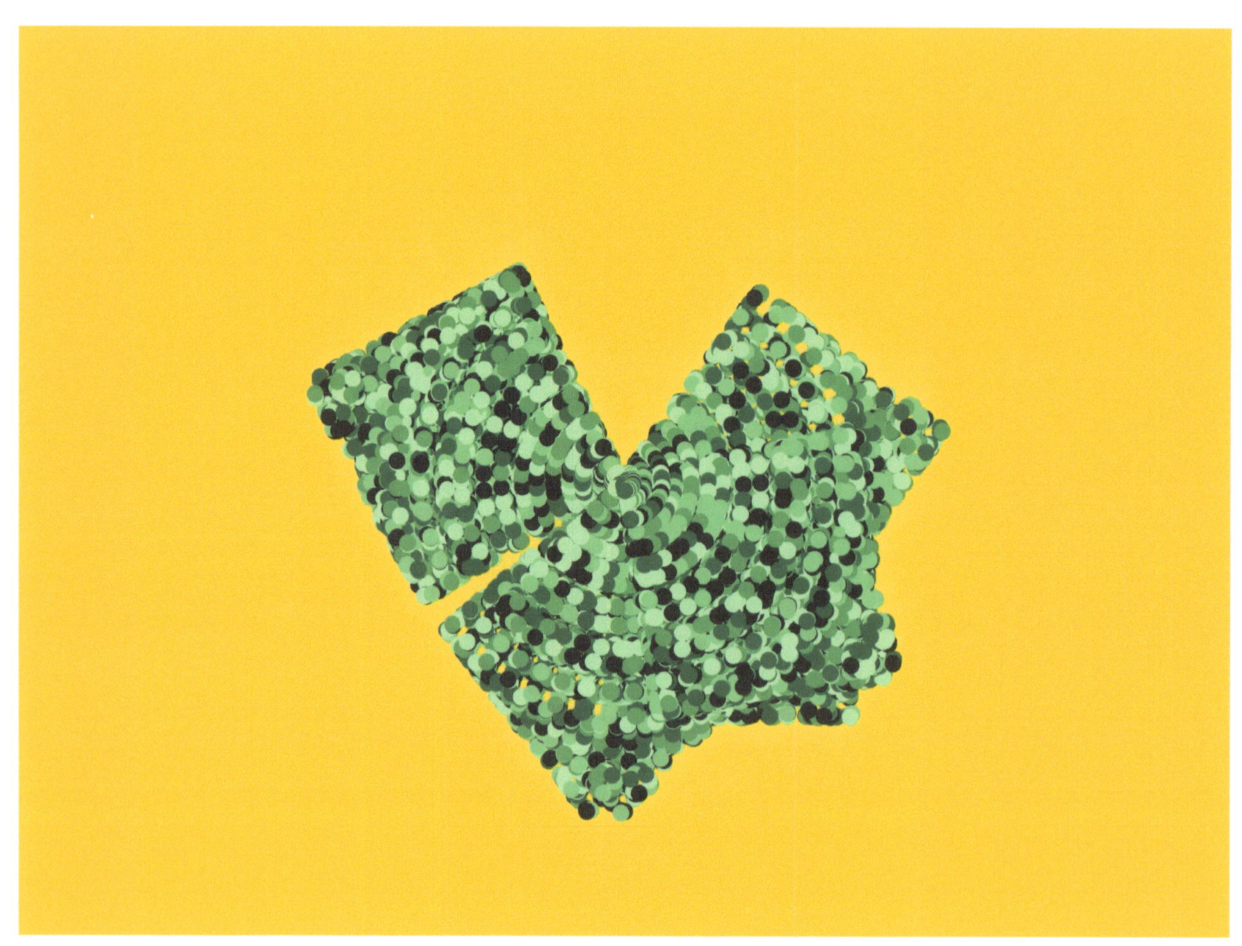

Figure 4.21: رَجْ (Dawn)

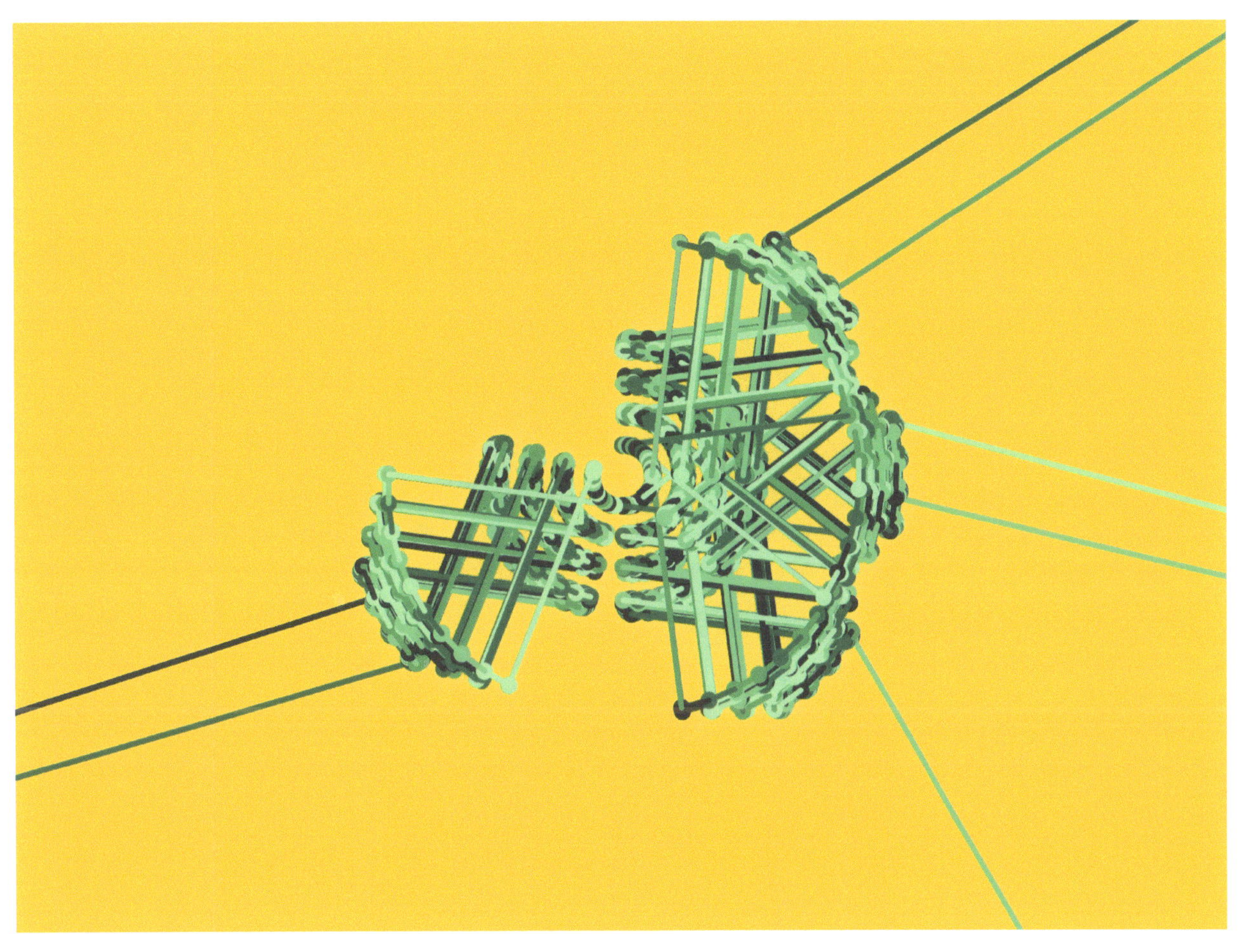

Figure 4.22: Melancolía (Melancholy)

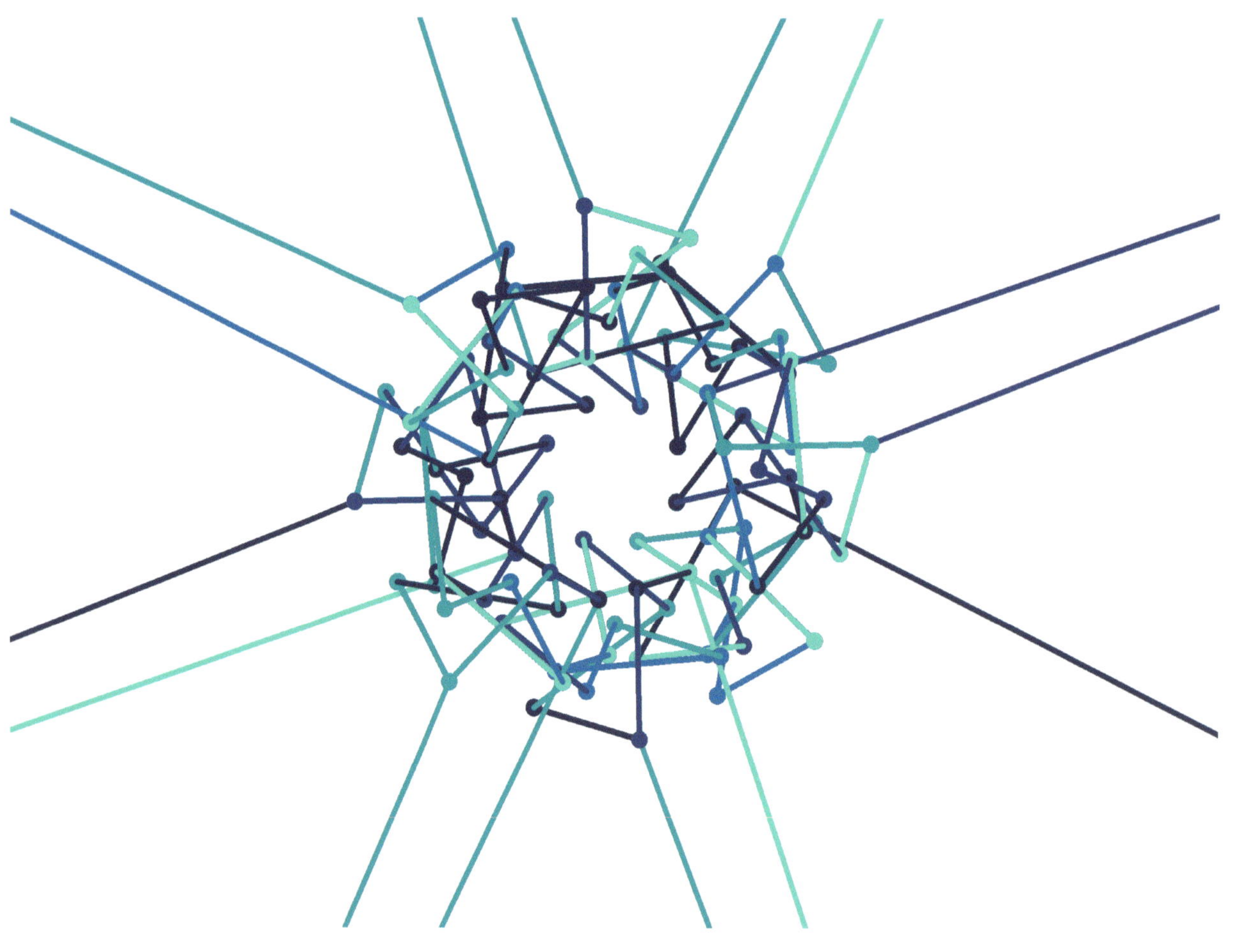

Figure 4.23: Ewigkeit (Eternity)

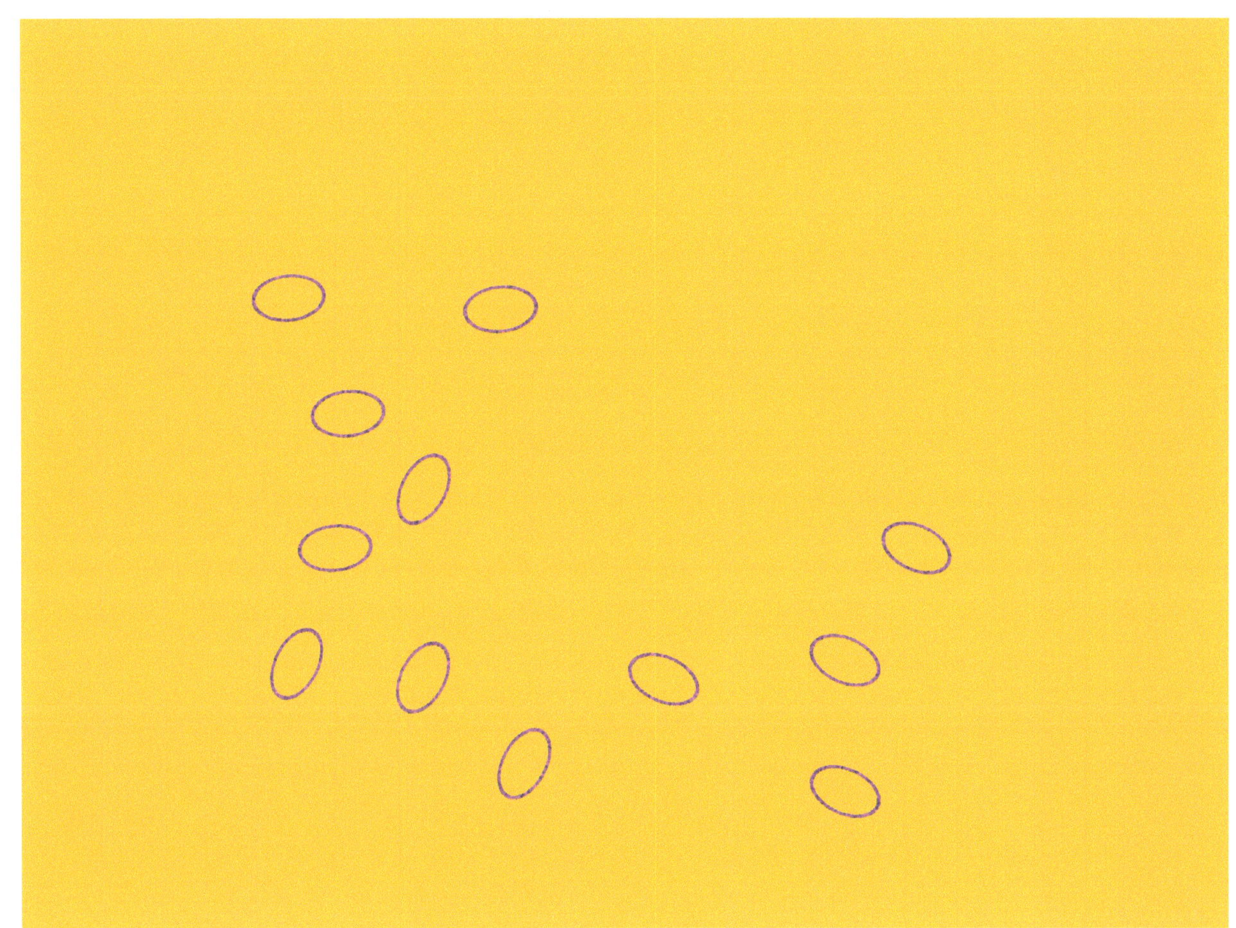

Figure 4.24: Destello (Sparkle)

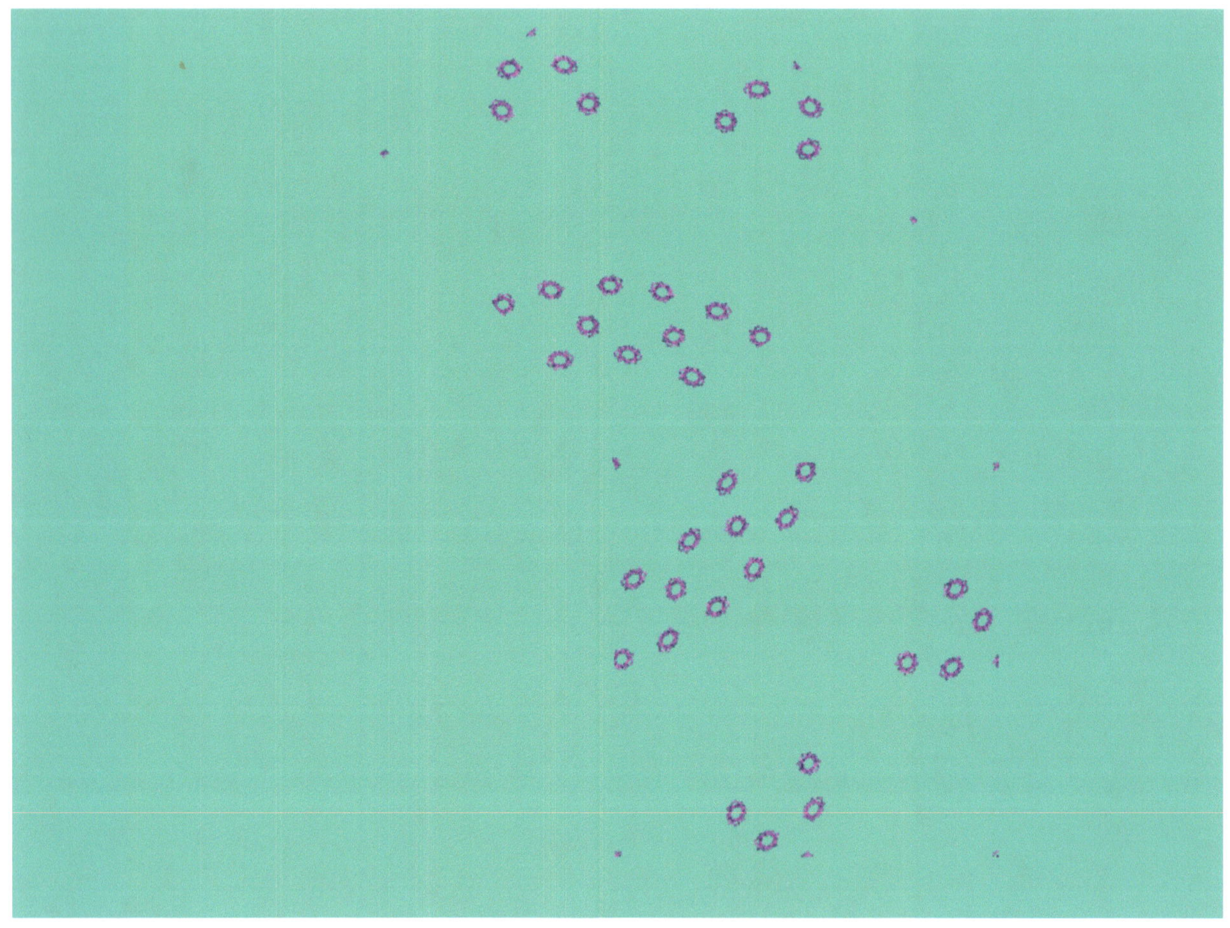

Figure 4.25: أمواج (Waves)

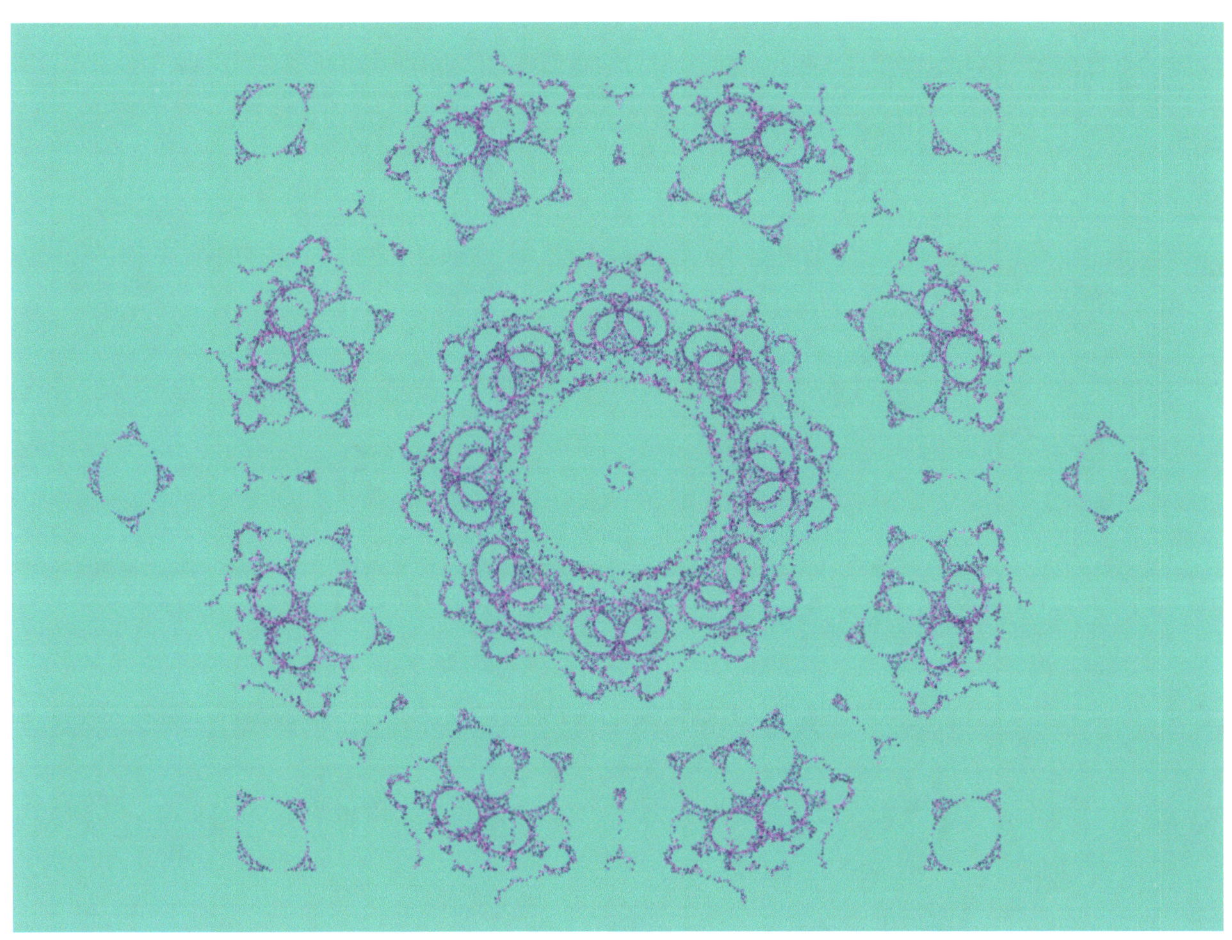

Figure 4.26: Sehnsucht (Longing)

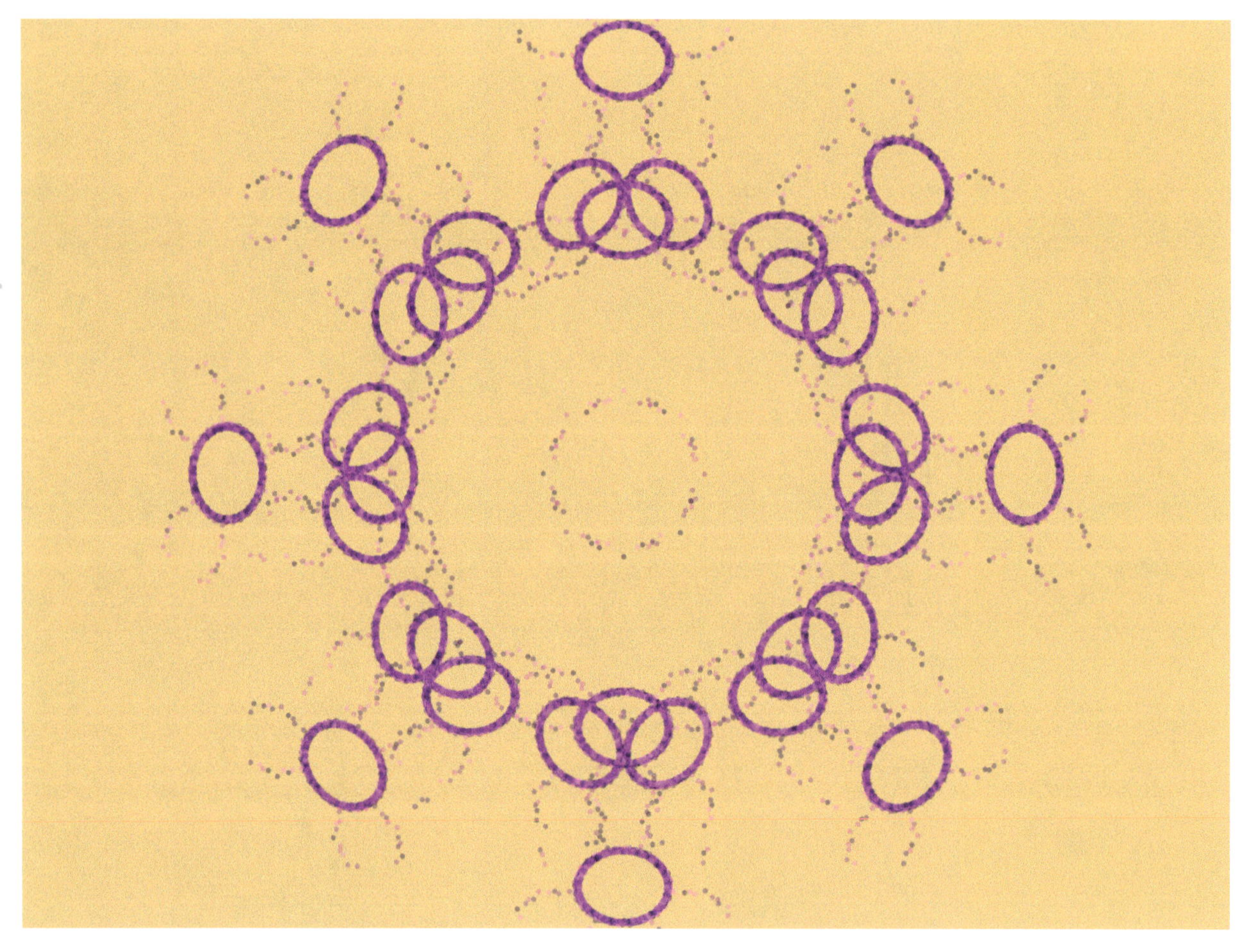

Figure 4.27: Inspiración (Inspiration)

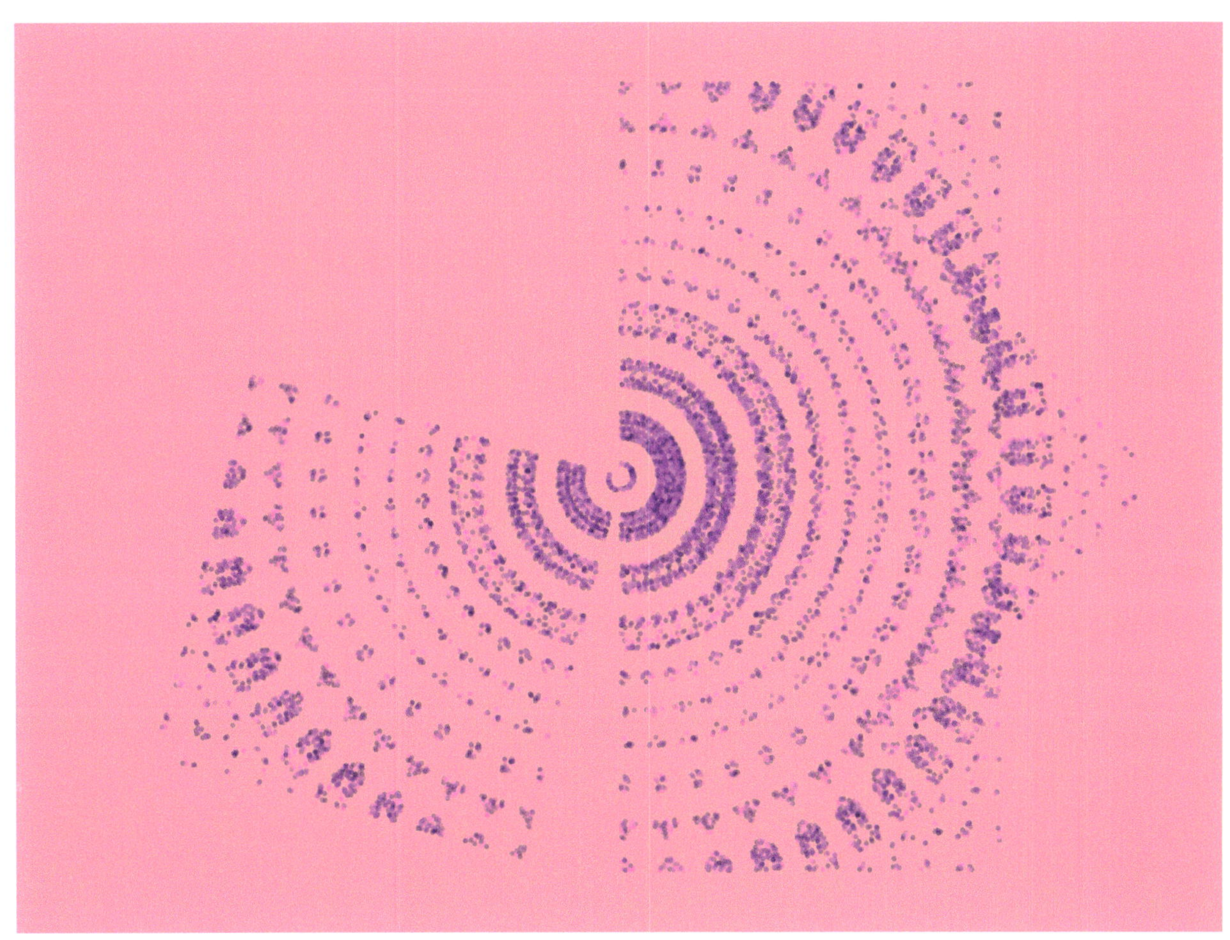

Figure 4.28: غرب (Twilight)

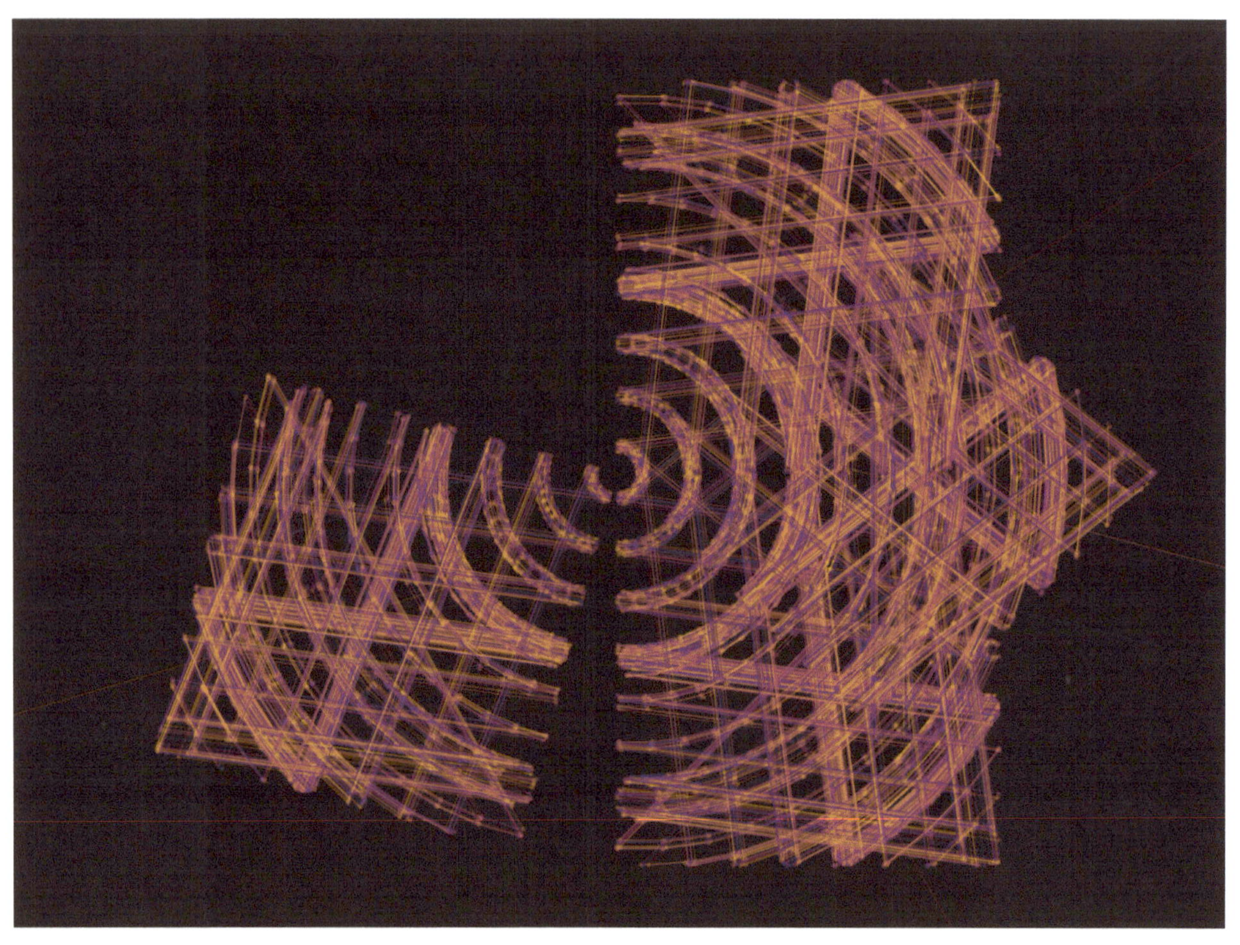

Figure 4.29: Lebensfreude (Joy of Life)

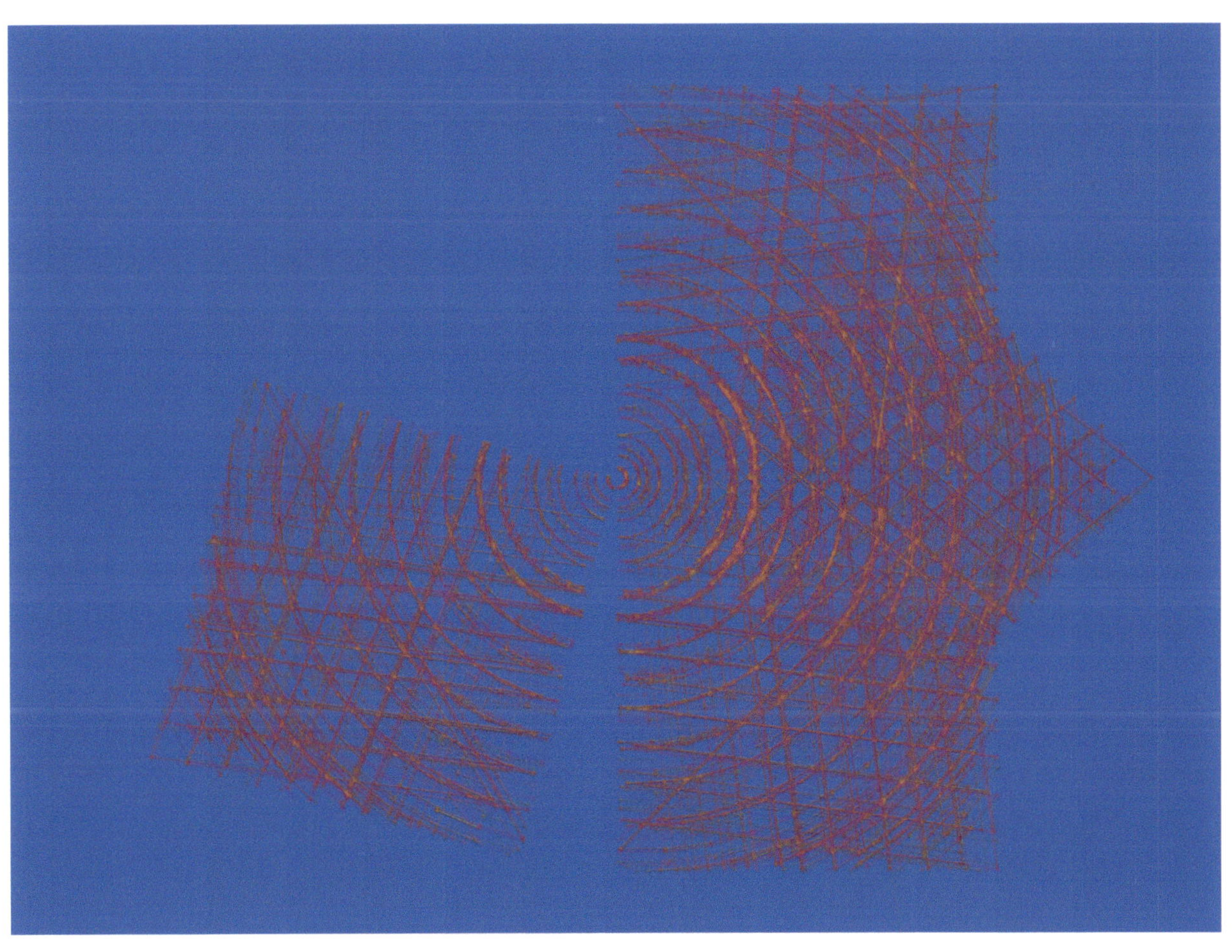

Figure 4.30: Nostalgia (Nostalgia)

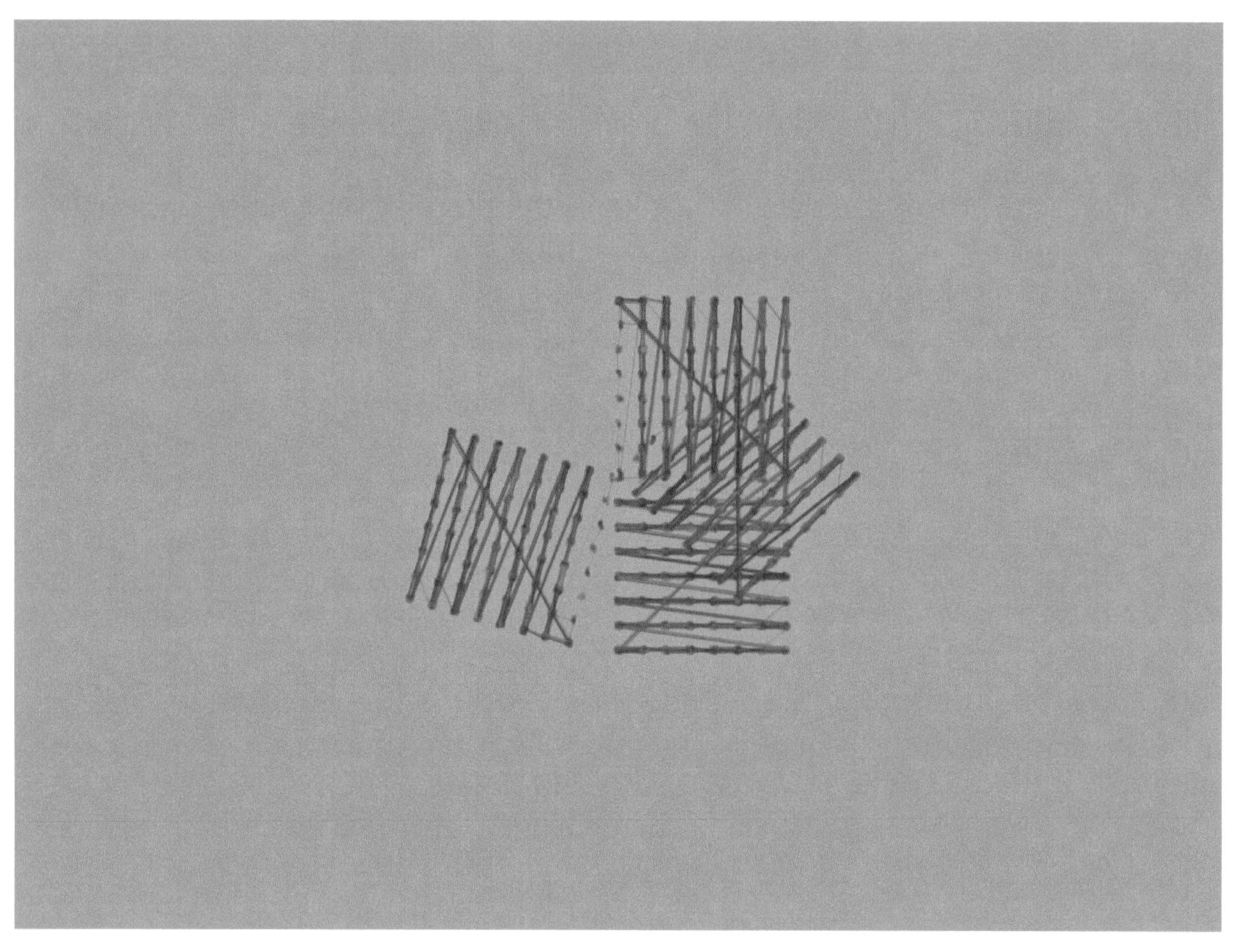

Figure 4.31: تأمل (Contemplation)

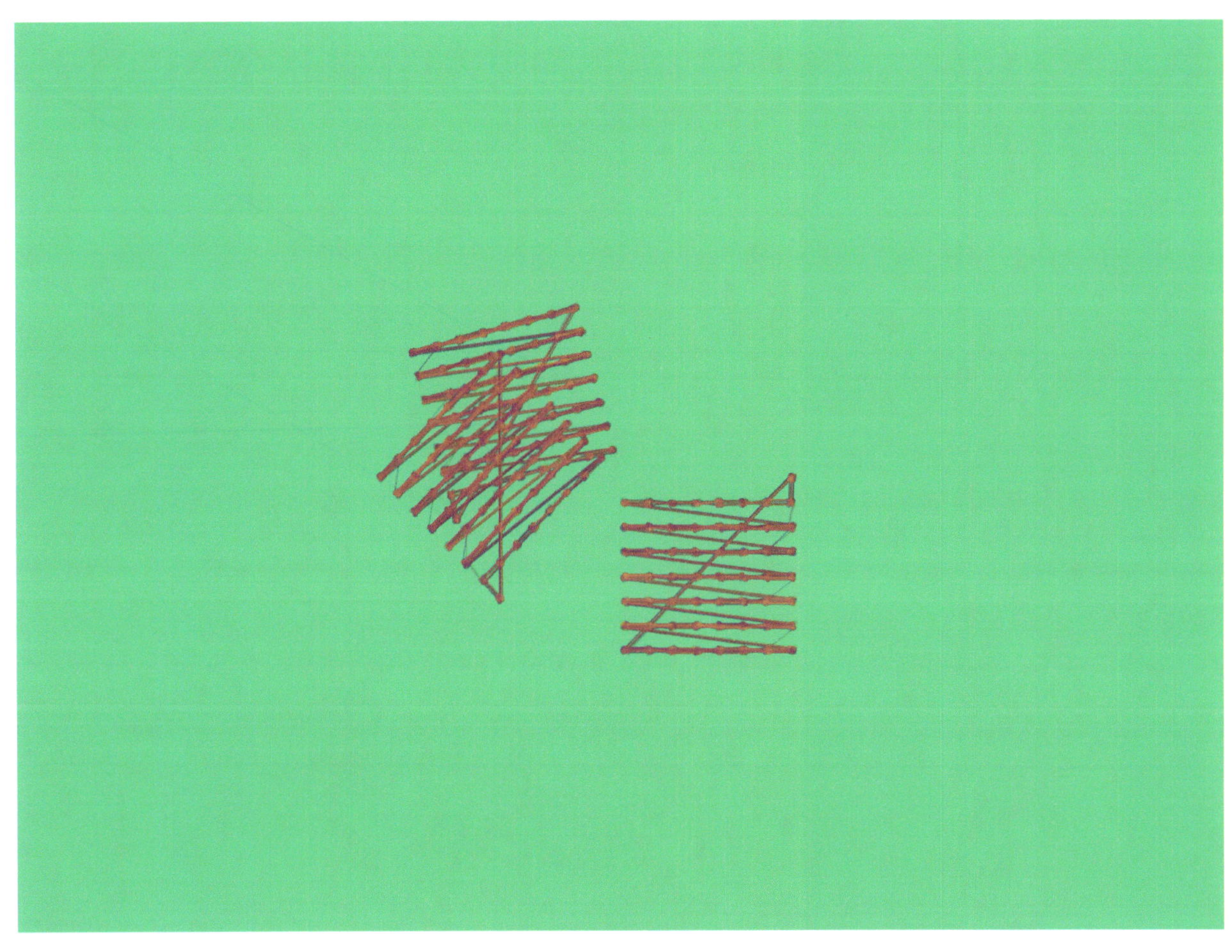

Figure 4.32: Erinnerung (Memory)

Figure 4.33: Brisa (Breeze)

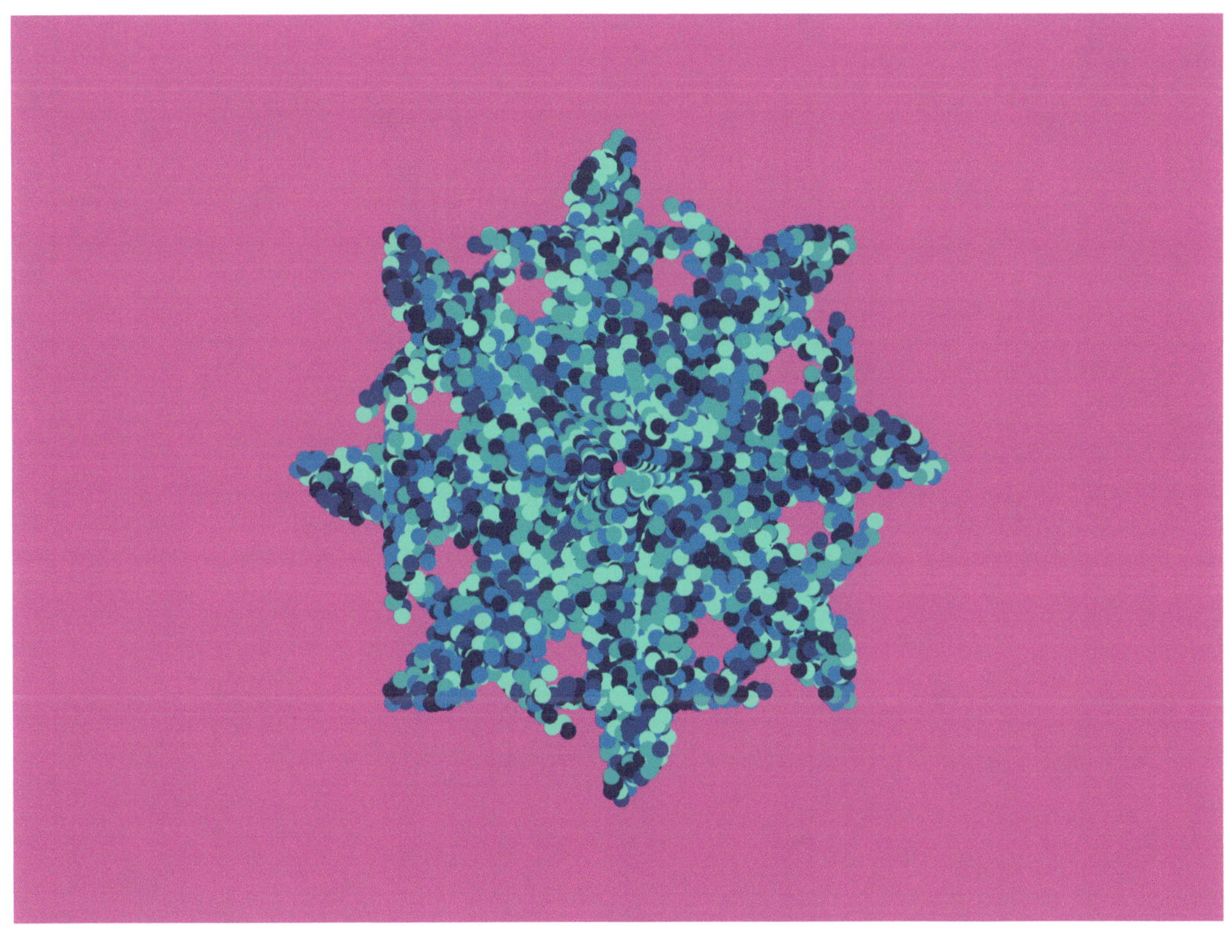

Figure 4.34: Entschlossenheit (Determination)

Figure 4.35: موسیقی (Music)

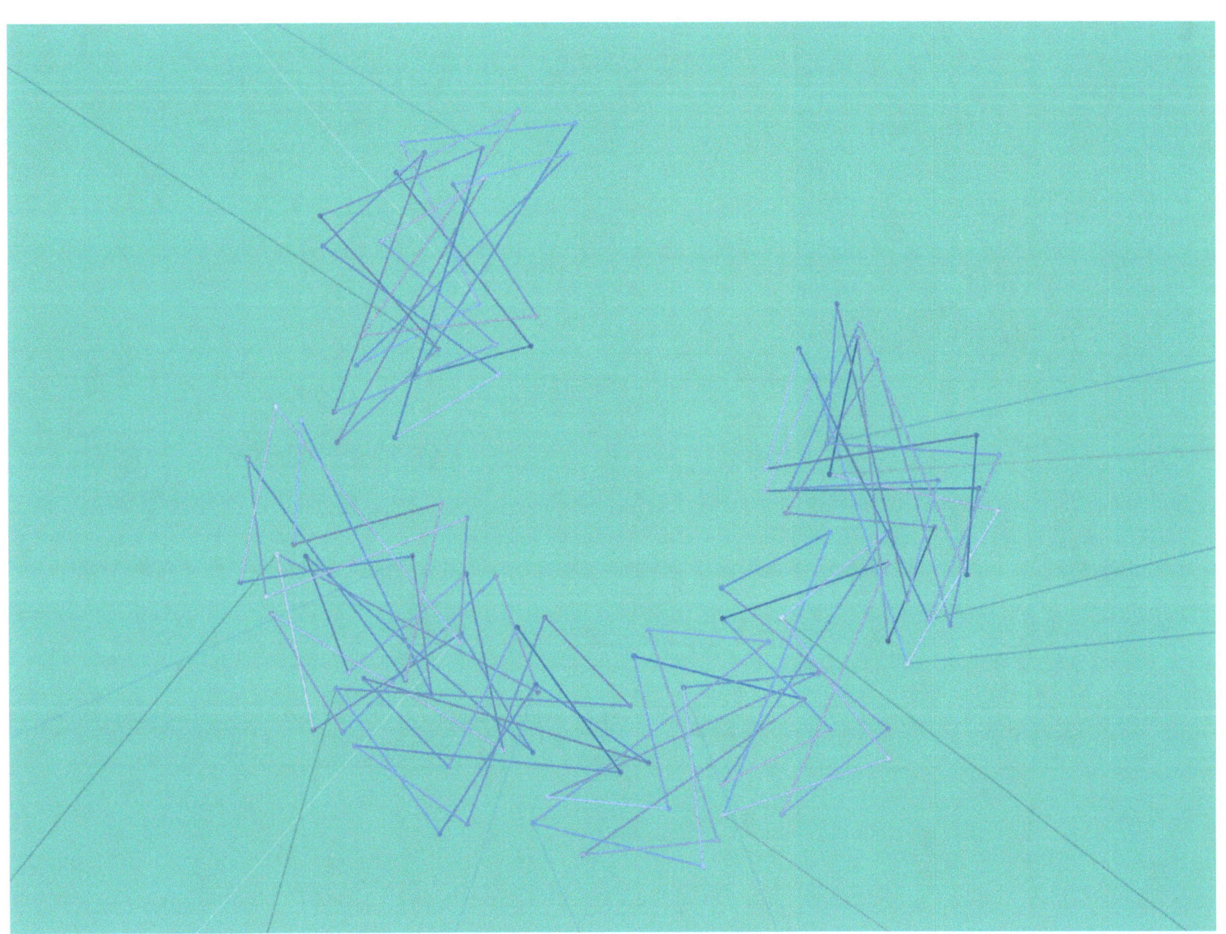

Figure 4.36: Intuición (Intuition)

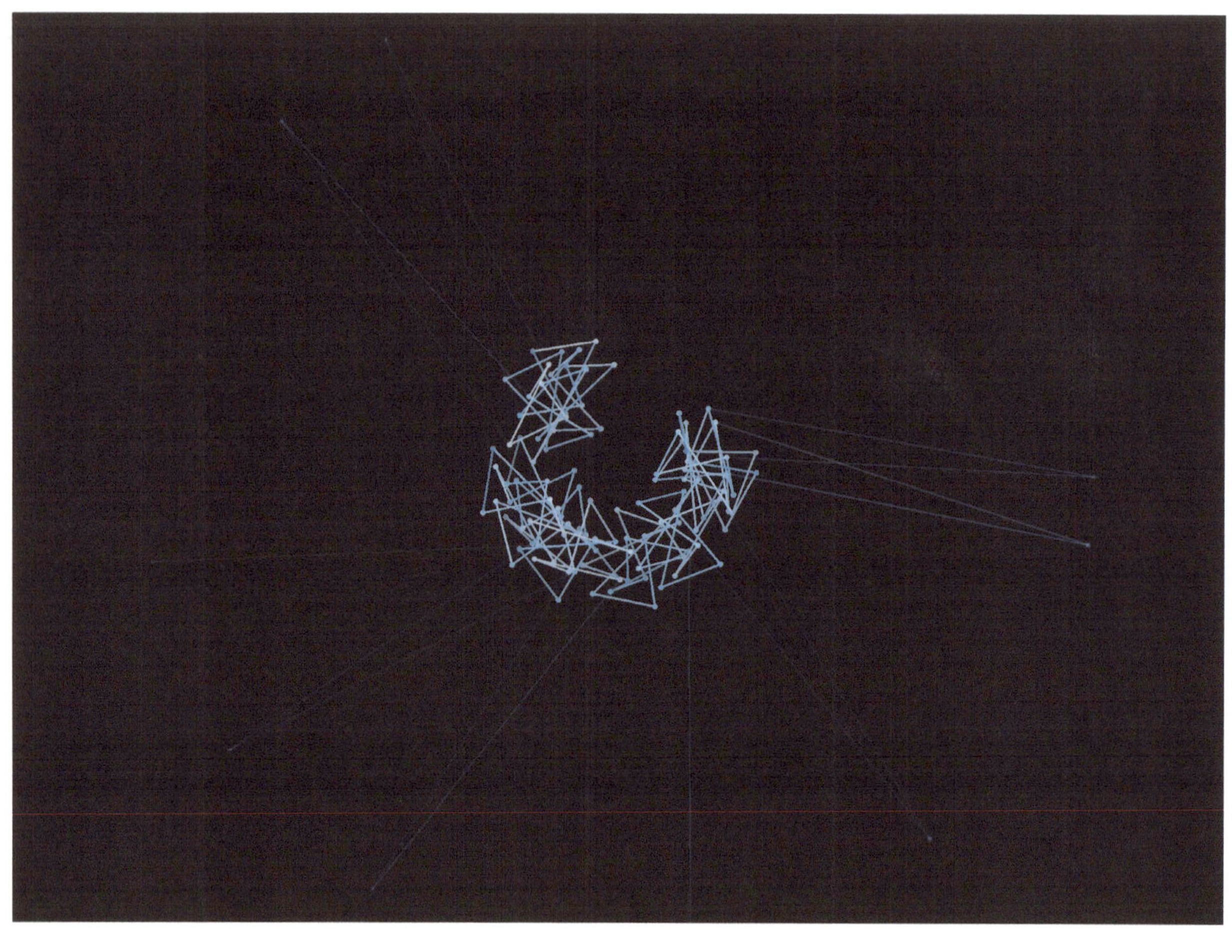

Figure 4.37: Geborgenheit (Security)

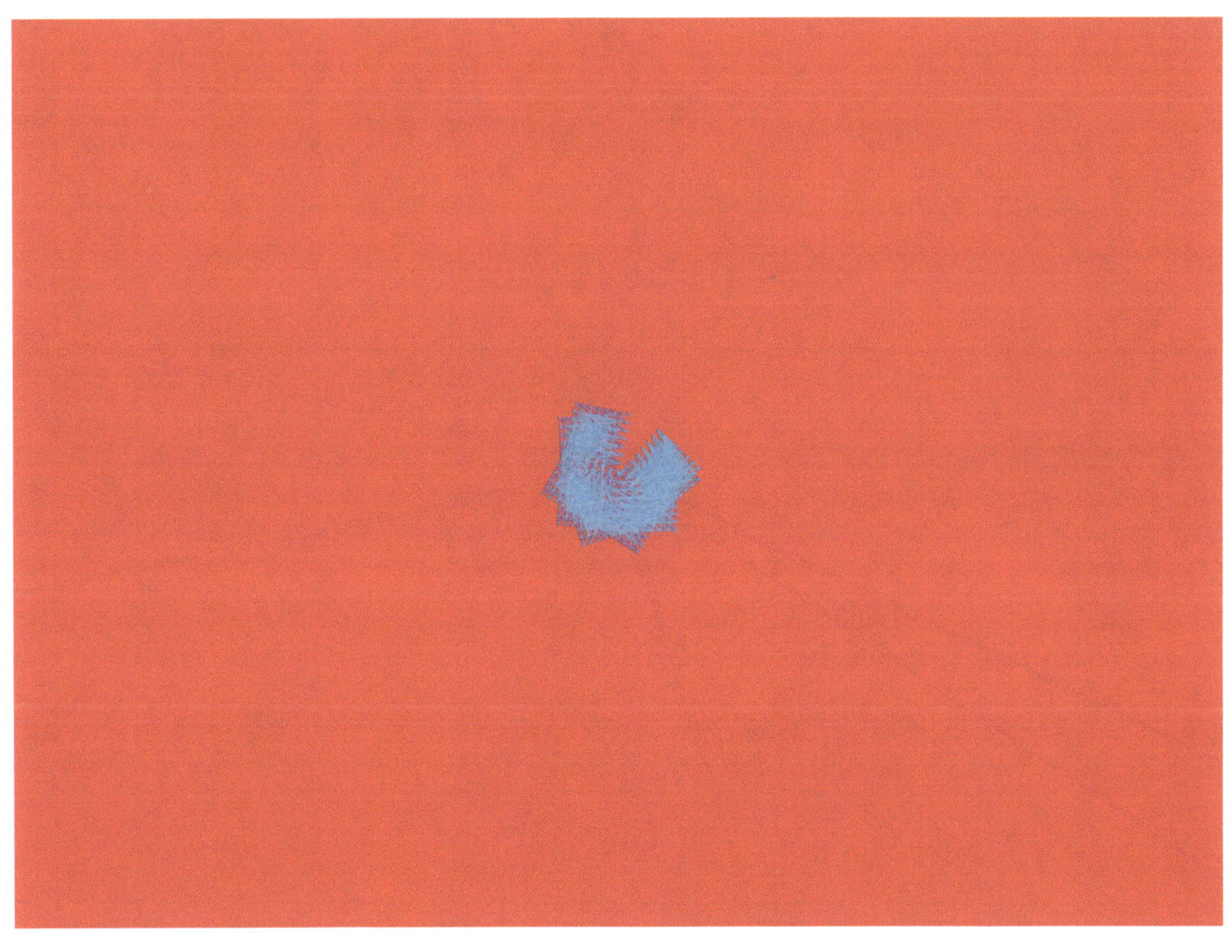

Figure 4.38: يحو (Inspiration)

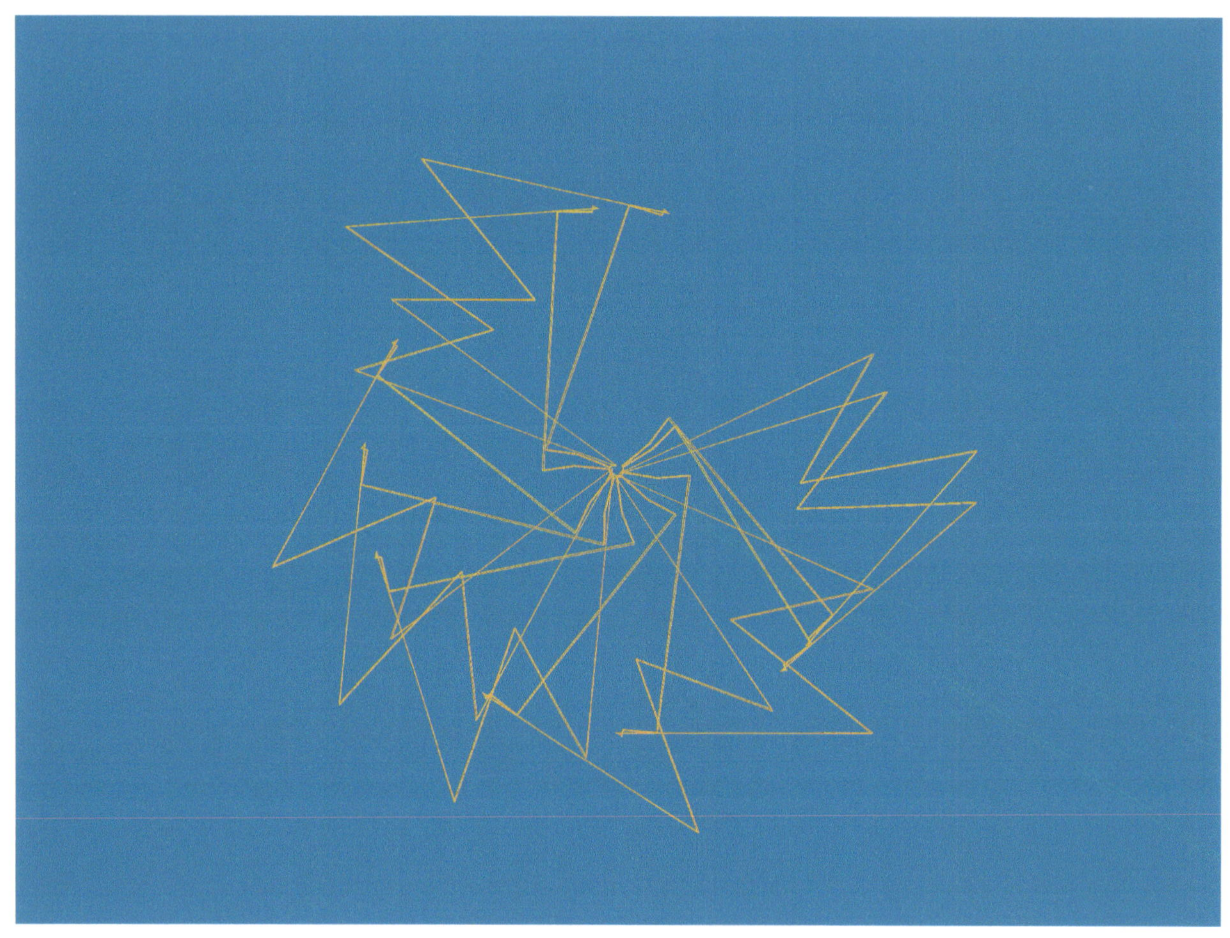

Figure 4.39: Claroscuro (Chiaroscuro)

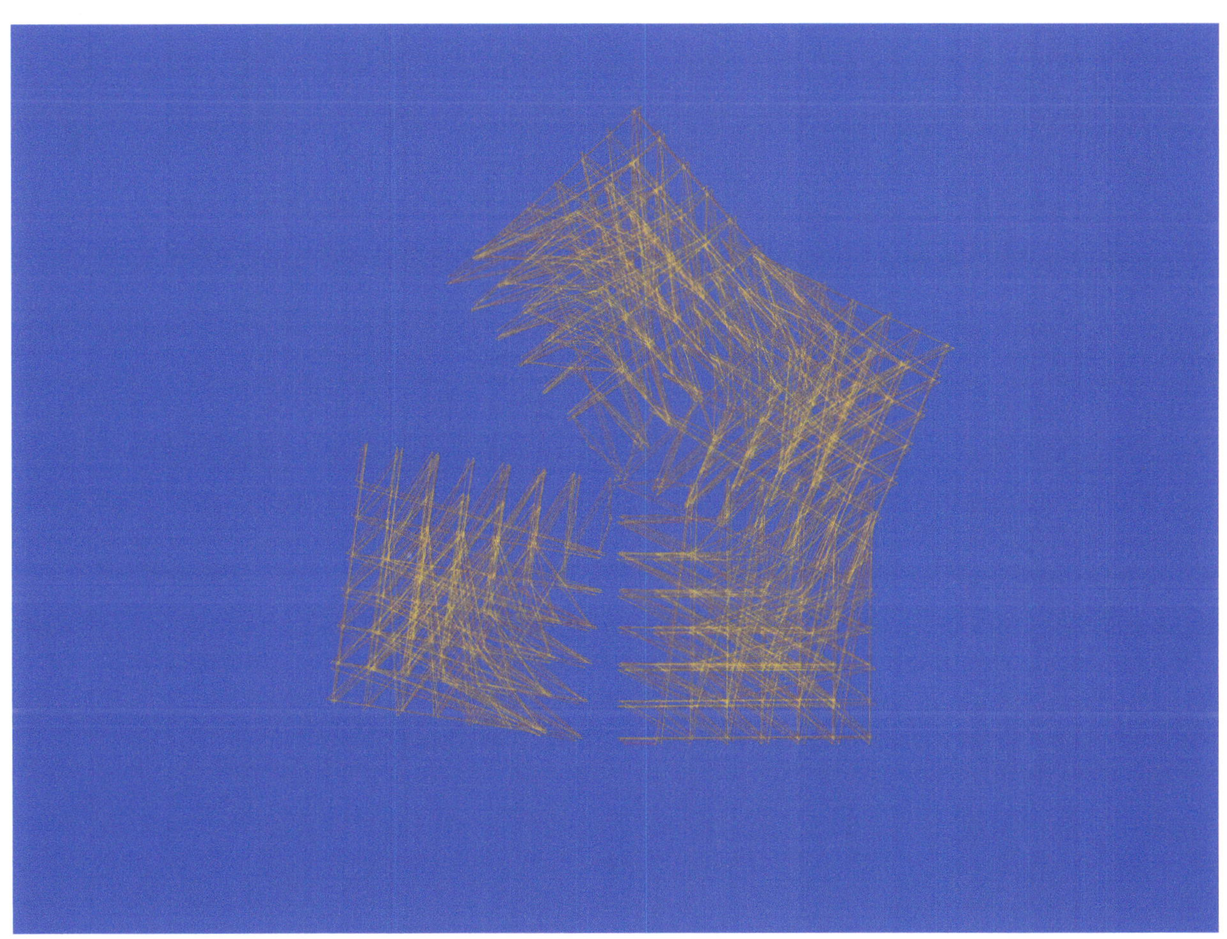

Figure 4.40: Zeitlos (Timeless)

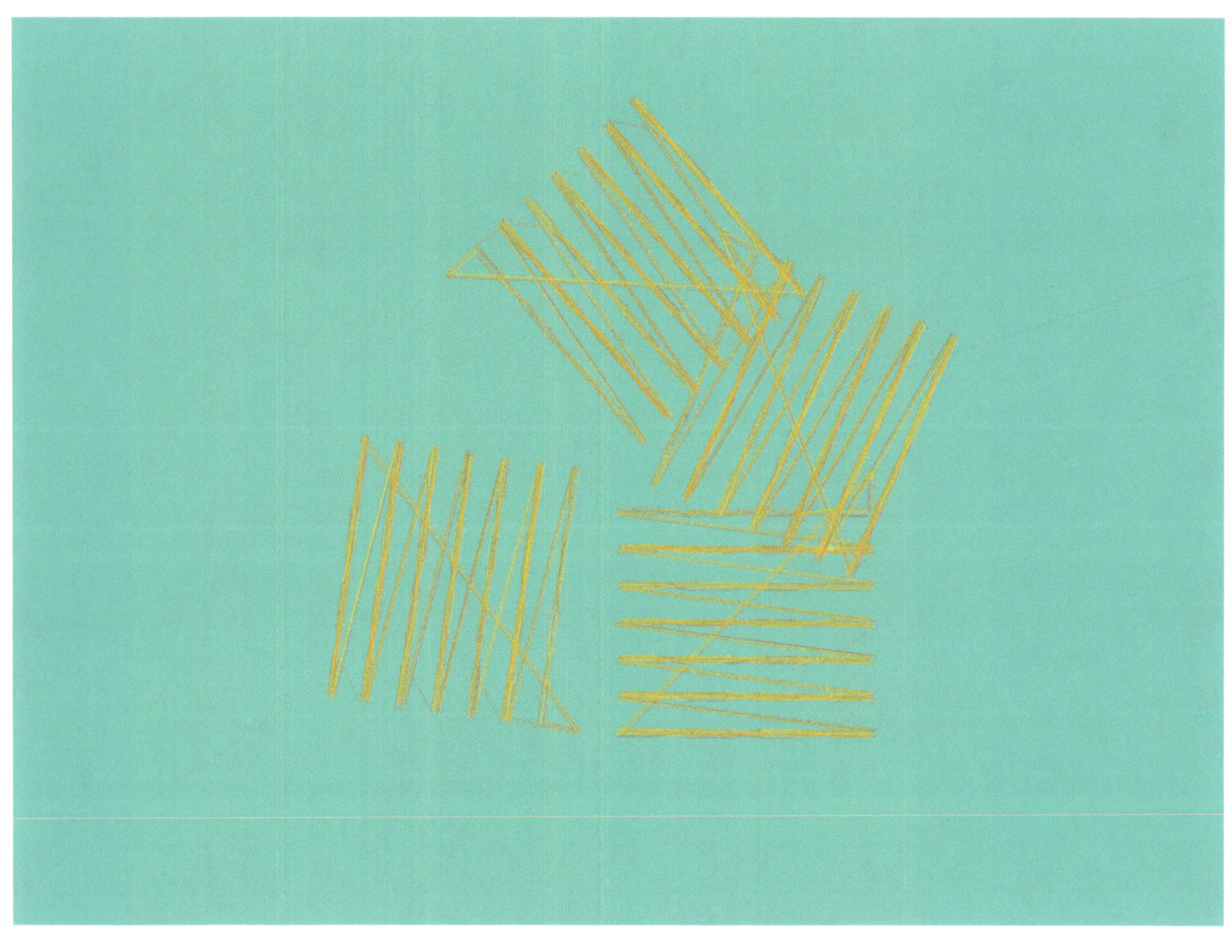

Figure 4.41: هدوء (Calmness)

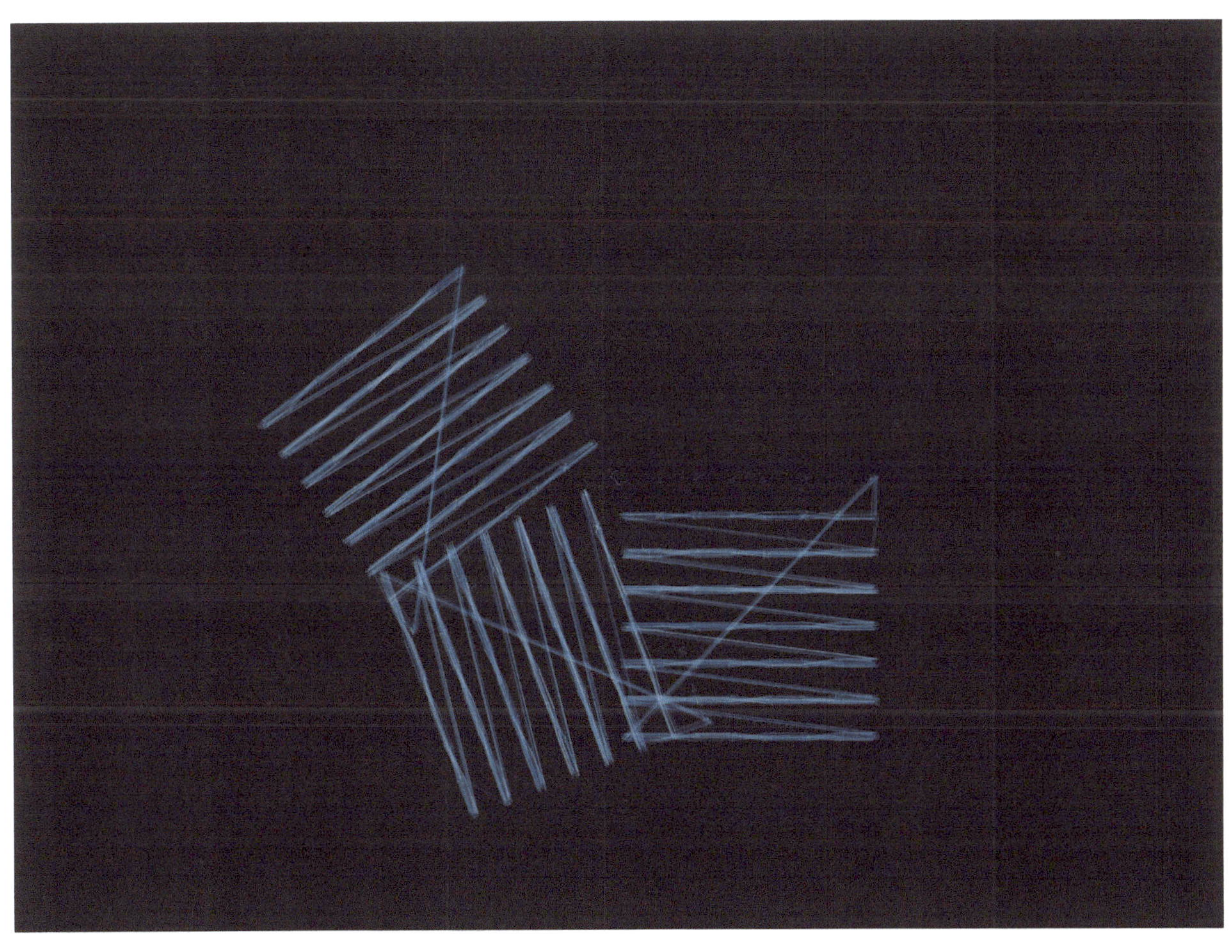

Figure 4.42: Cautiverio (Captivity)

Figure 4.43: Erforschung (Exploration)

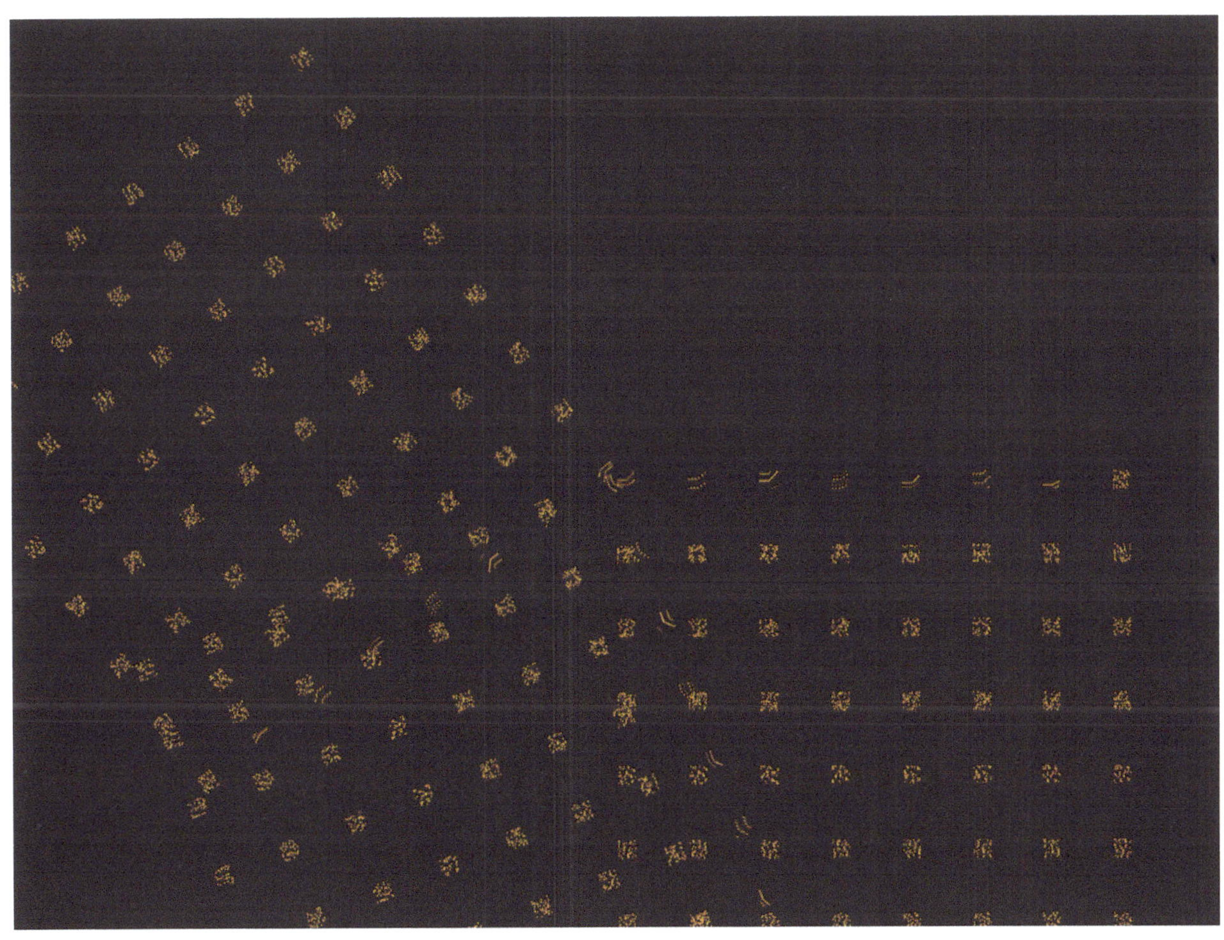

Figure 4.44: ابتسامة (Smile)

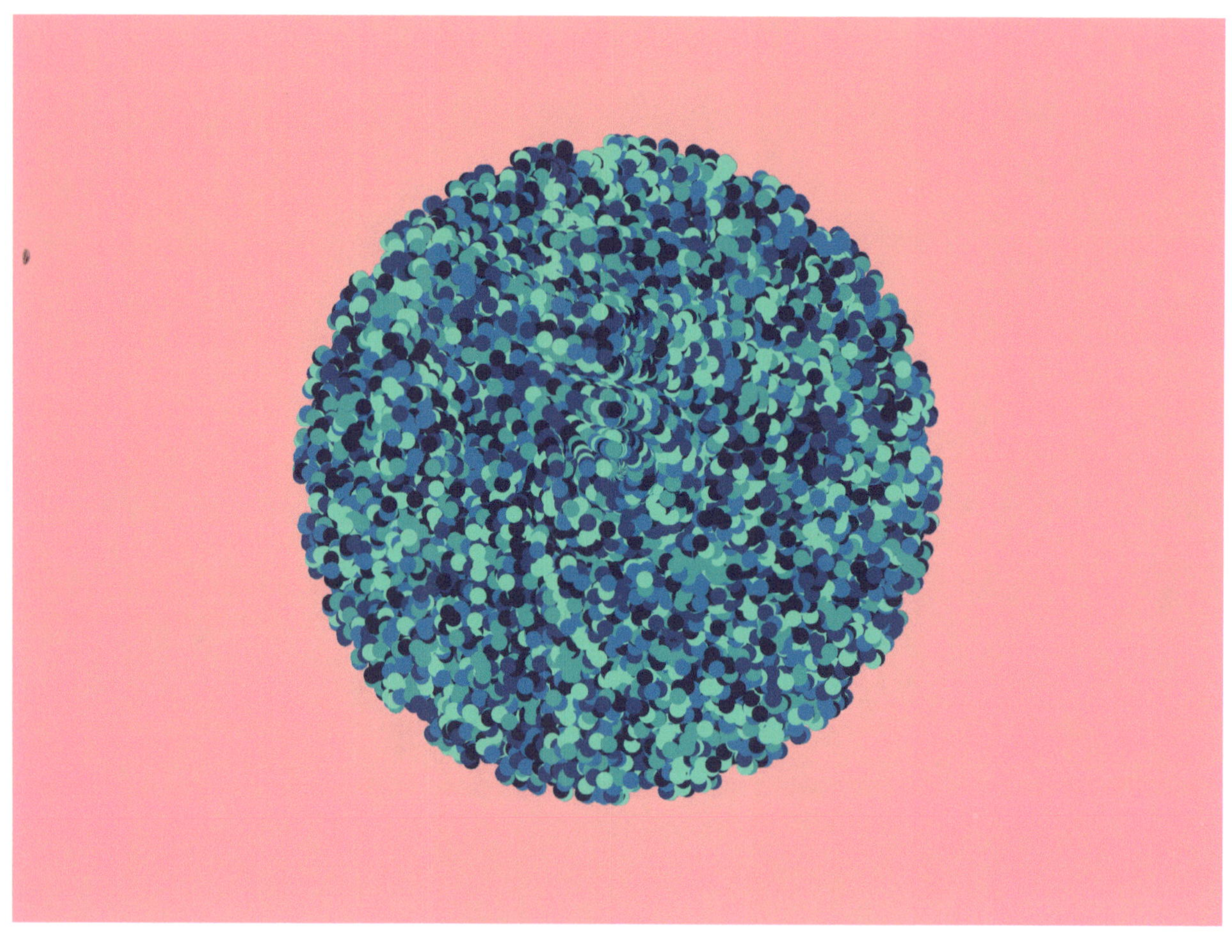

Figure 4.45: Amistad (Friendship)

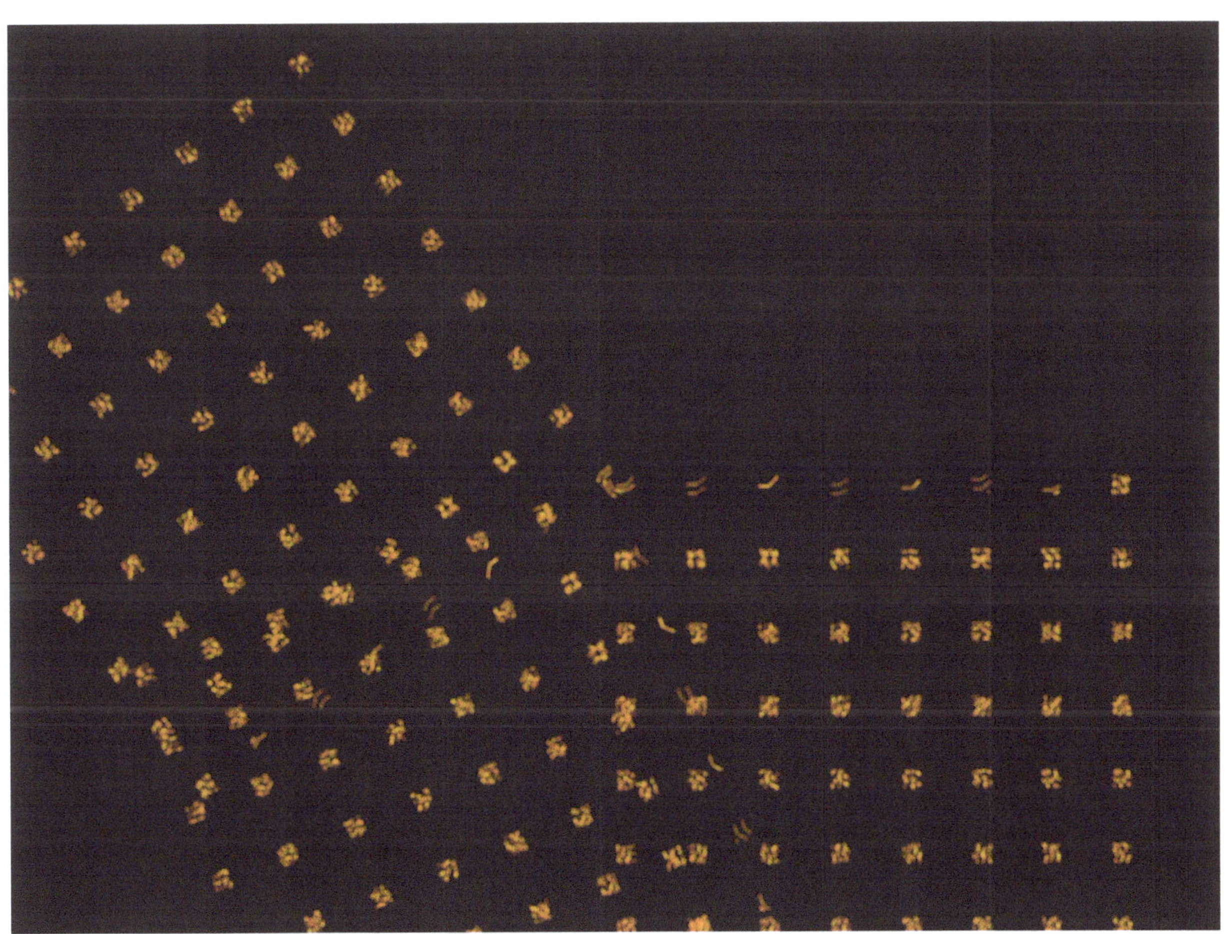

Figure 4.46: Wachstum (Growth)

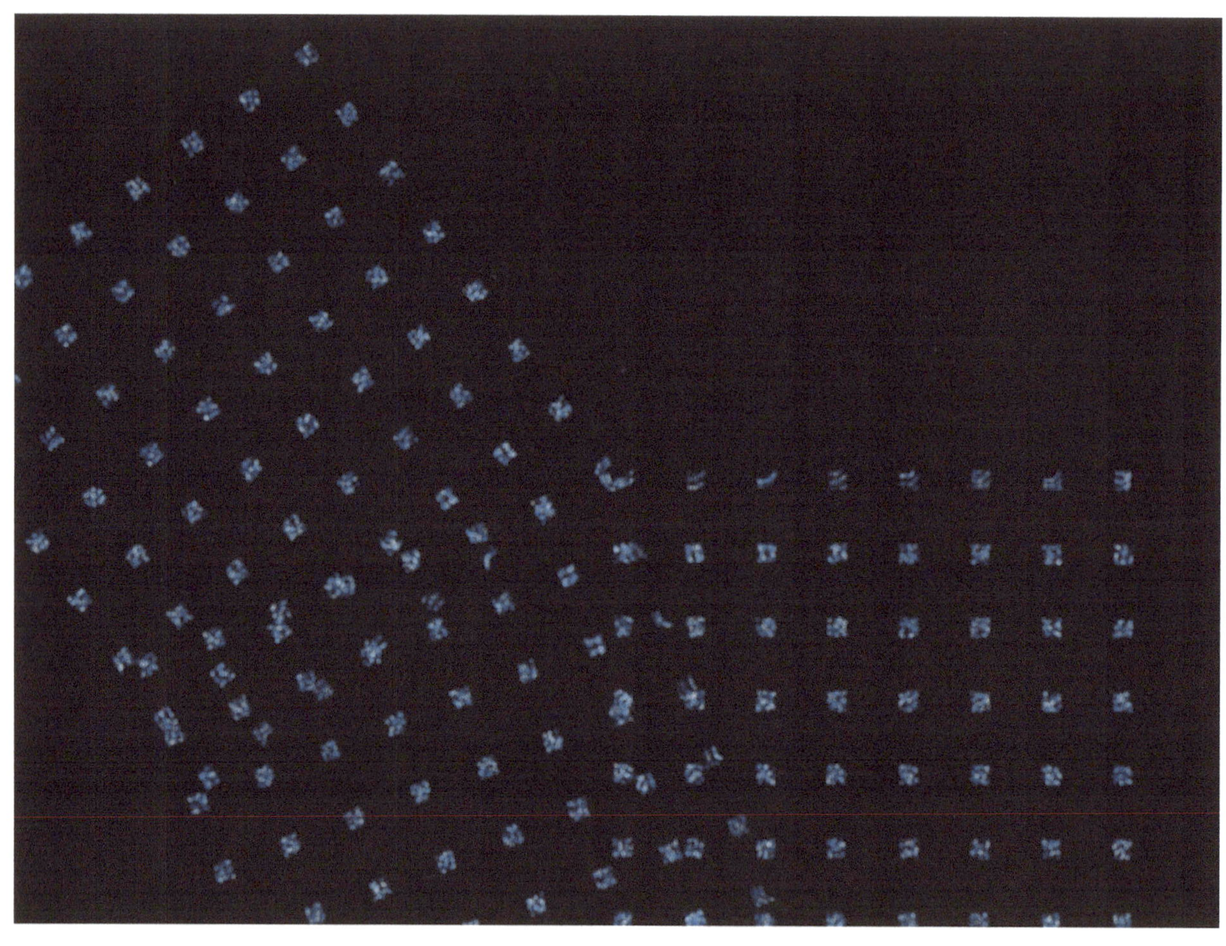

Figure 4.47: غموض (Mystery)

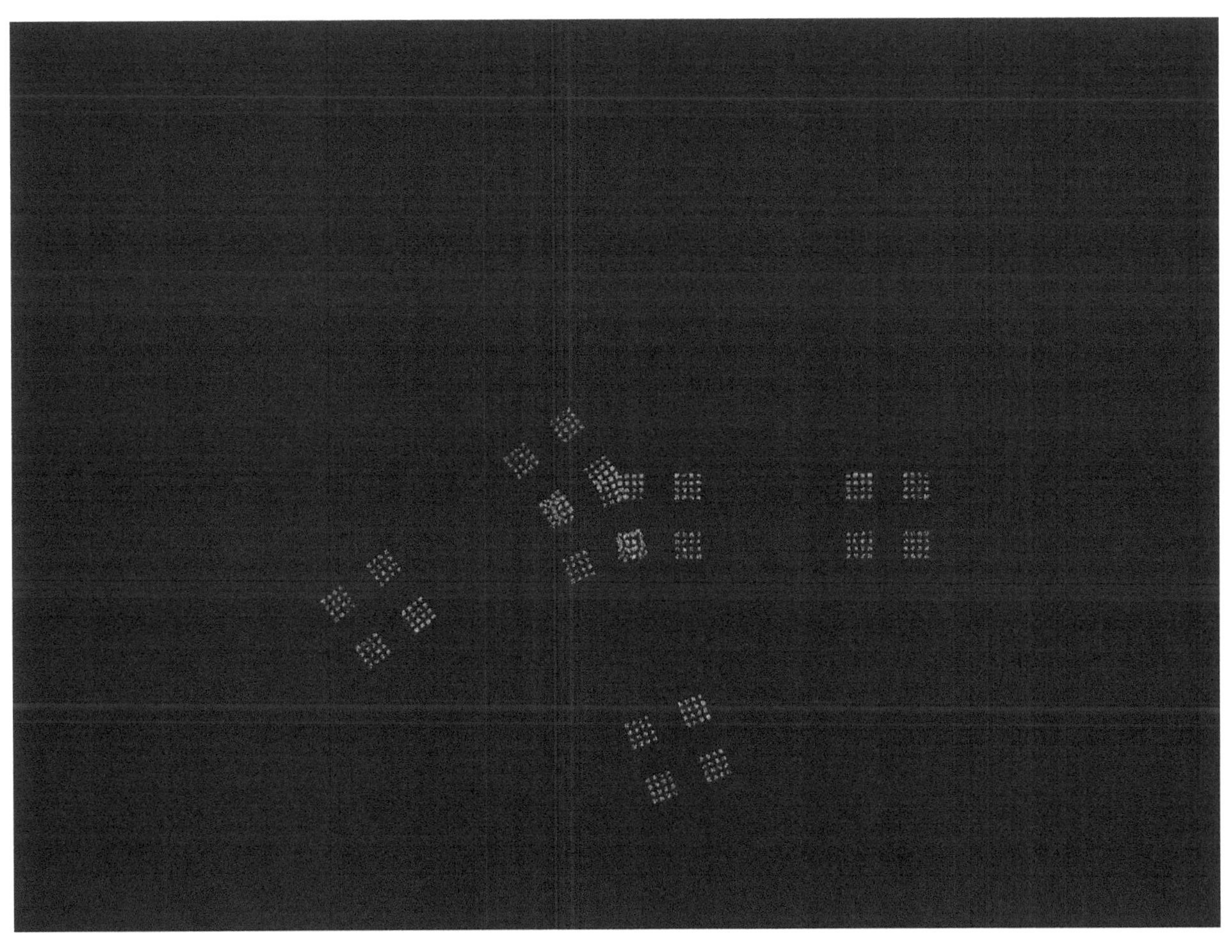

Figure 4.48: Renacimiento (Rebirth)

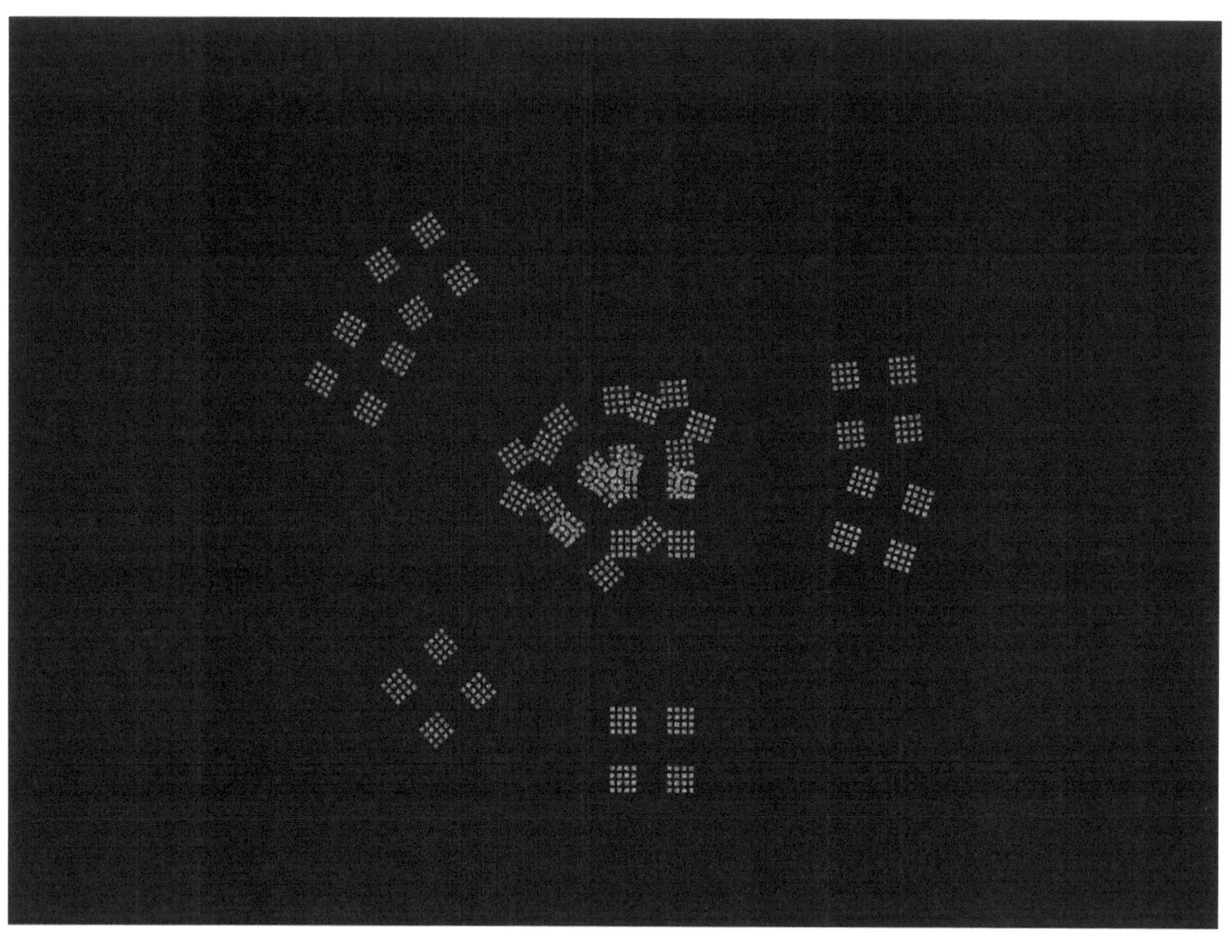

Figure 4.49: Vergänglichkeit (Transience)

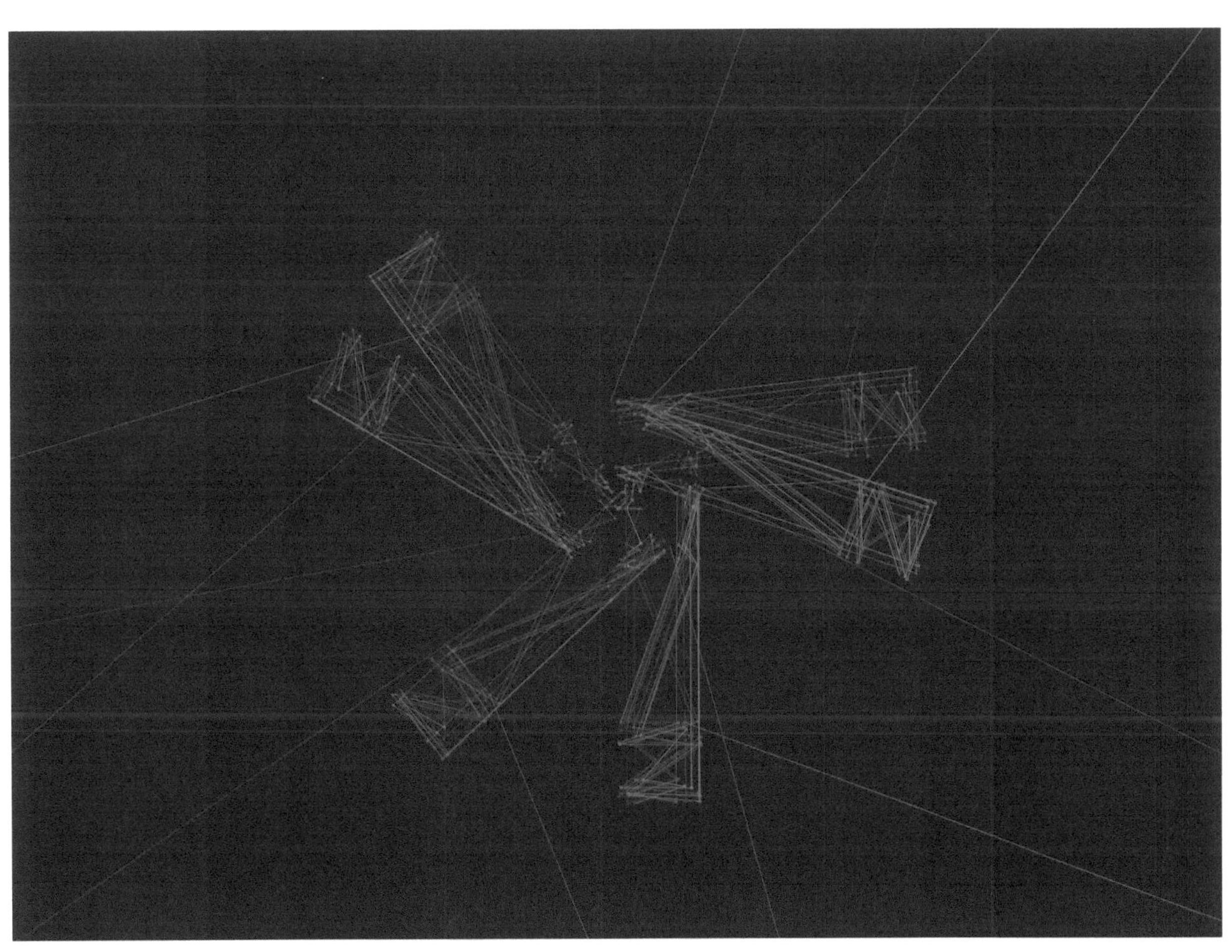

Figure 4.50: ألوان (Colors)

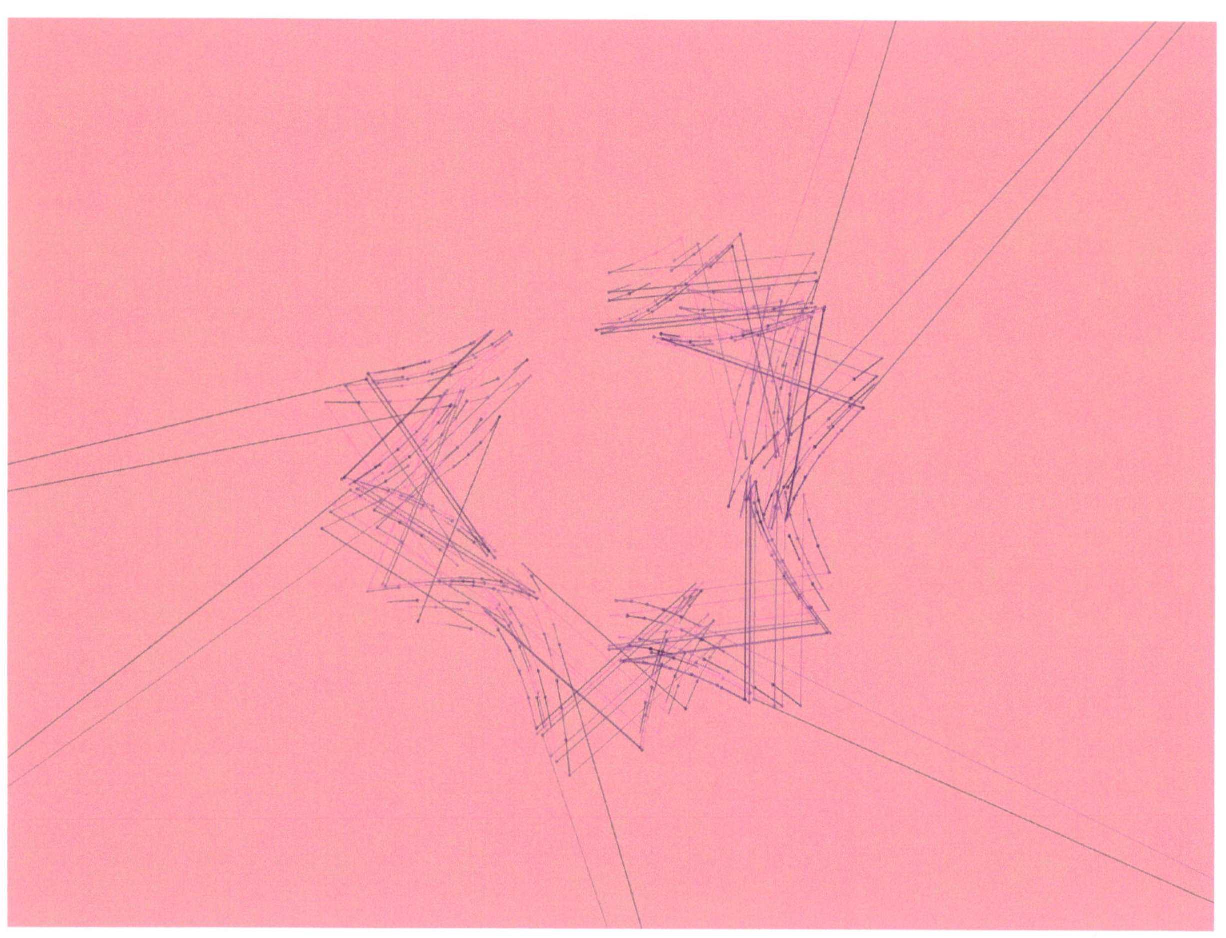

Figure 4.51: Esperanza (Hope)

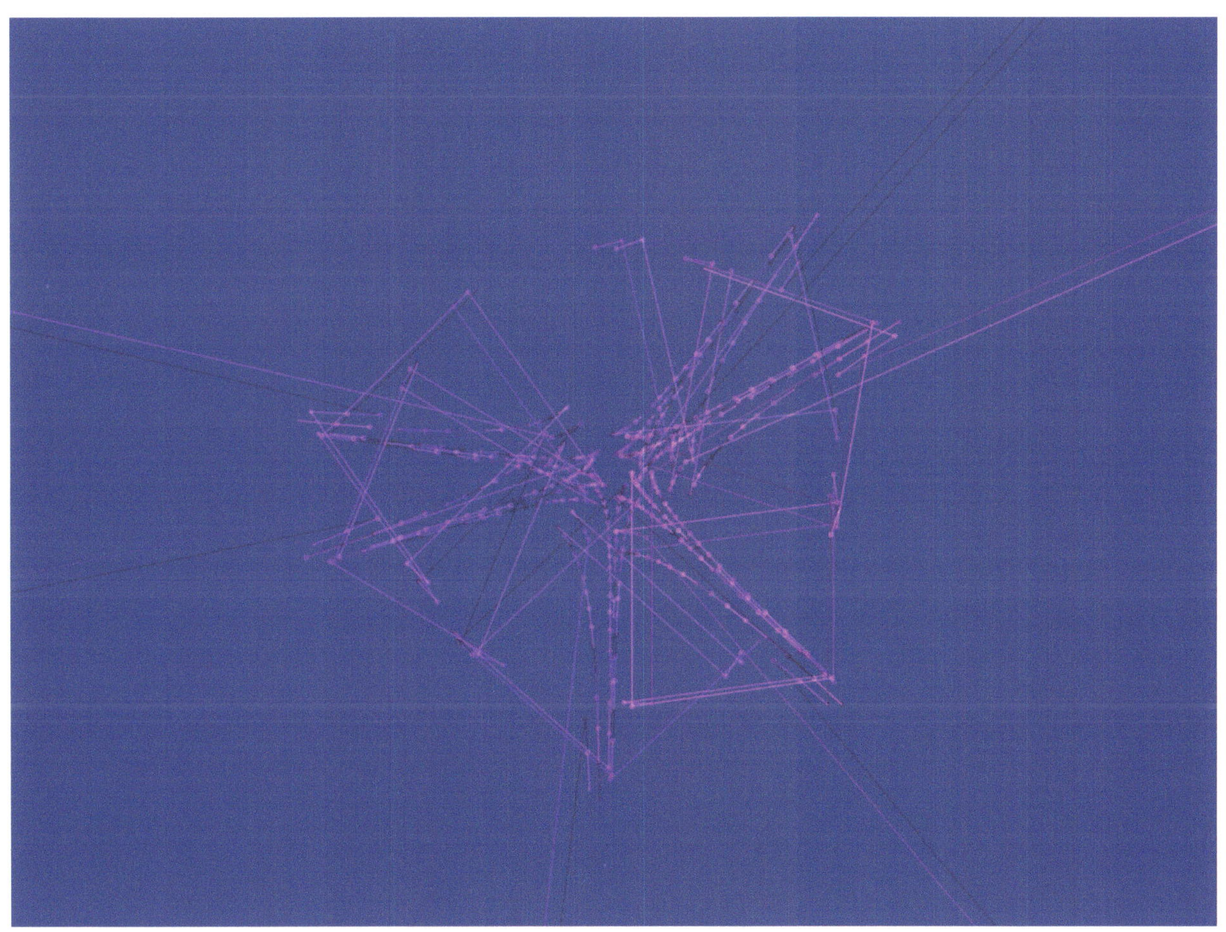

Figure 4.52: Verschmelzung (Fusion)

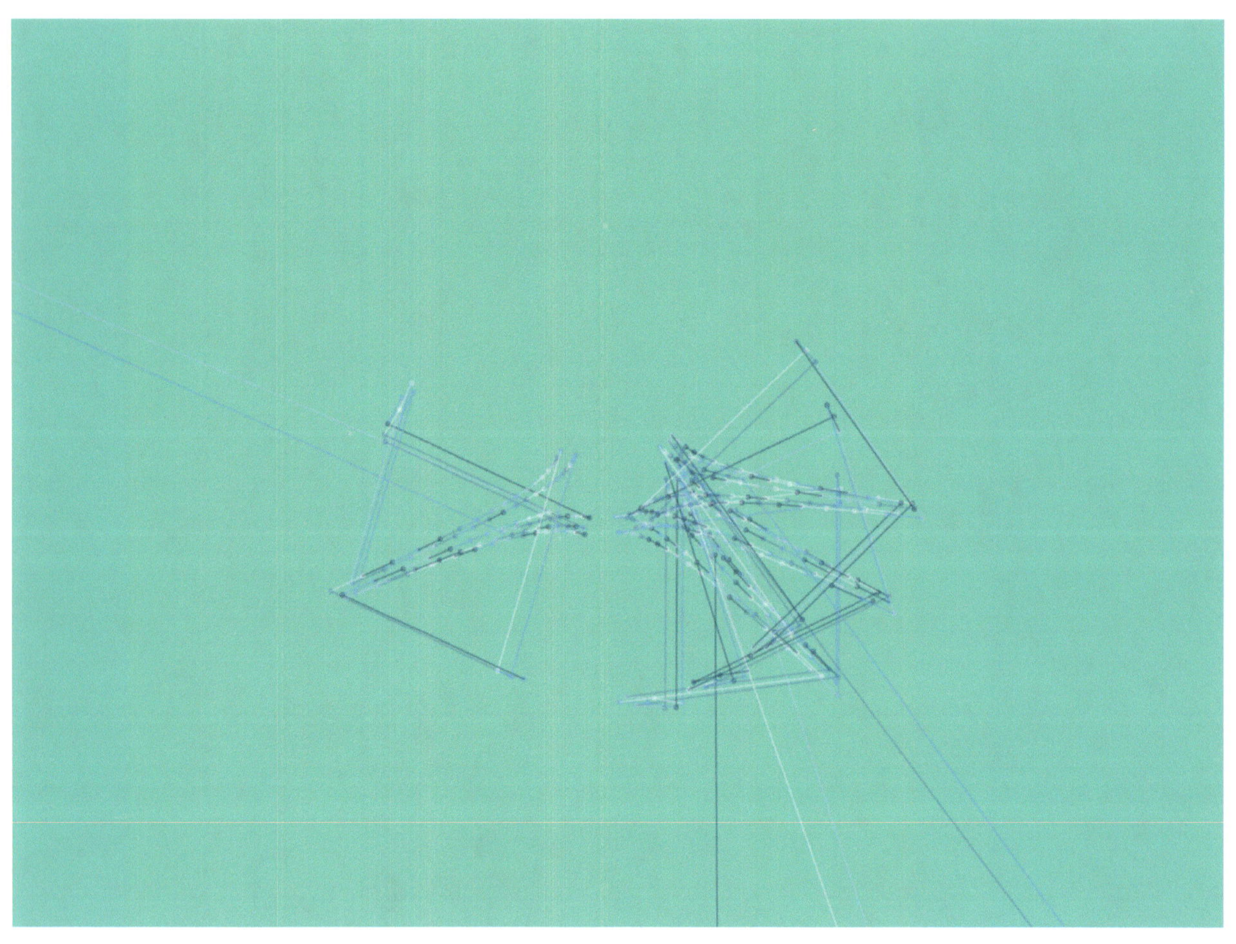

Figure 4.53: فرح (Joy)

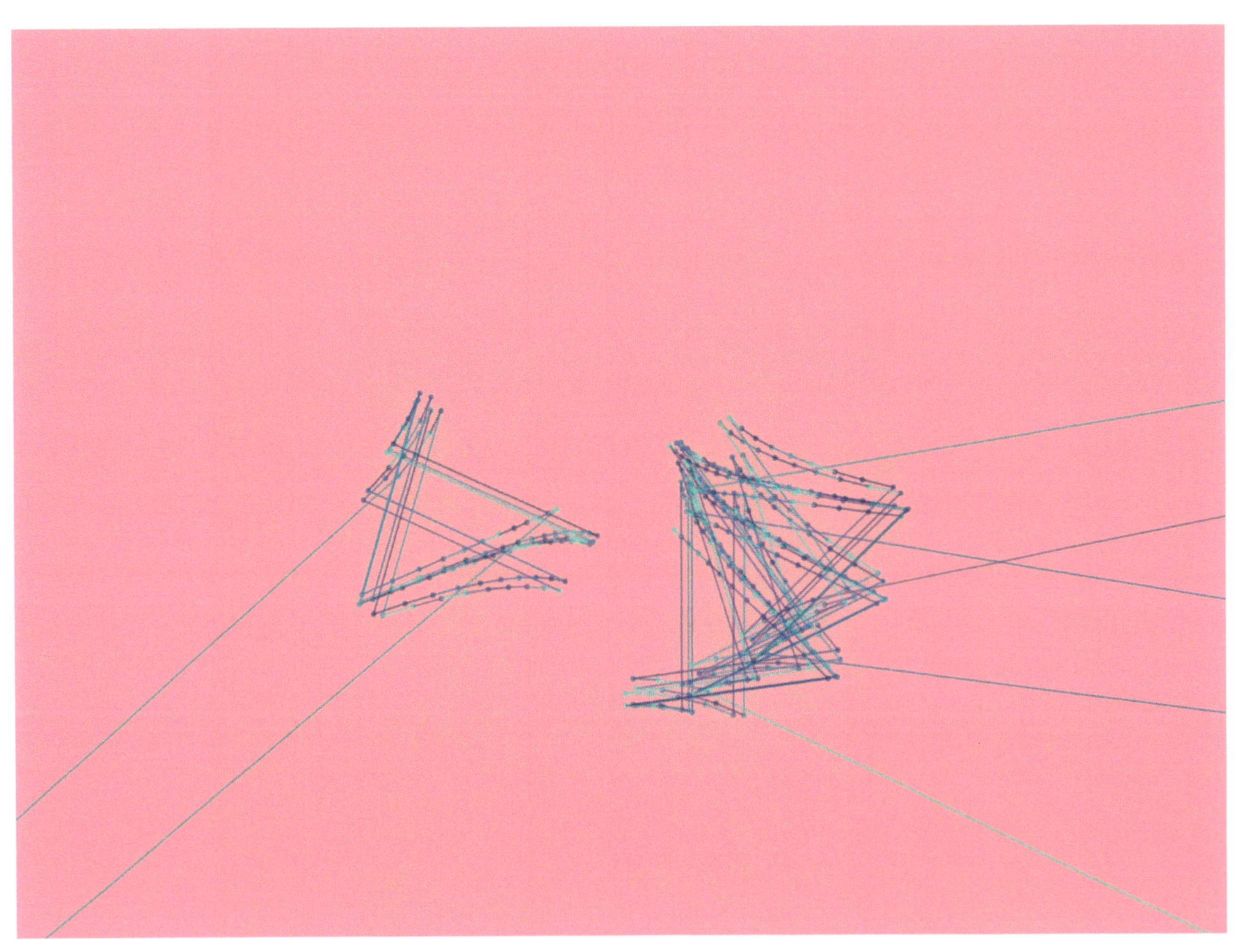

Figure 4.54: Armonía (Harmony)

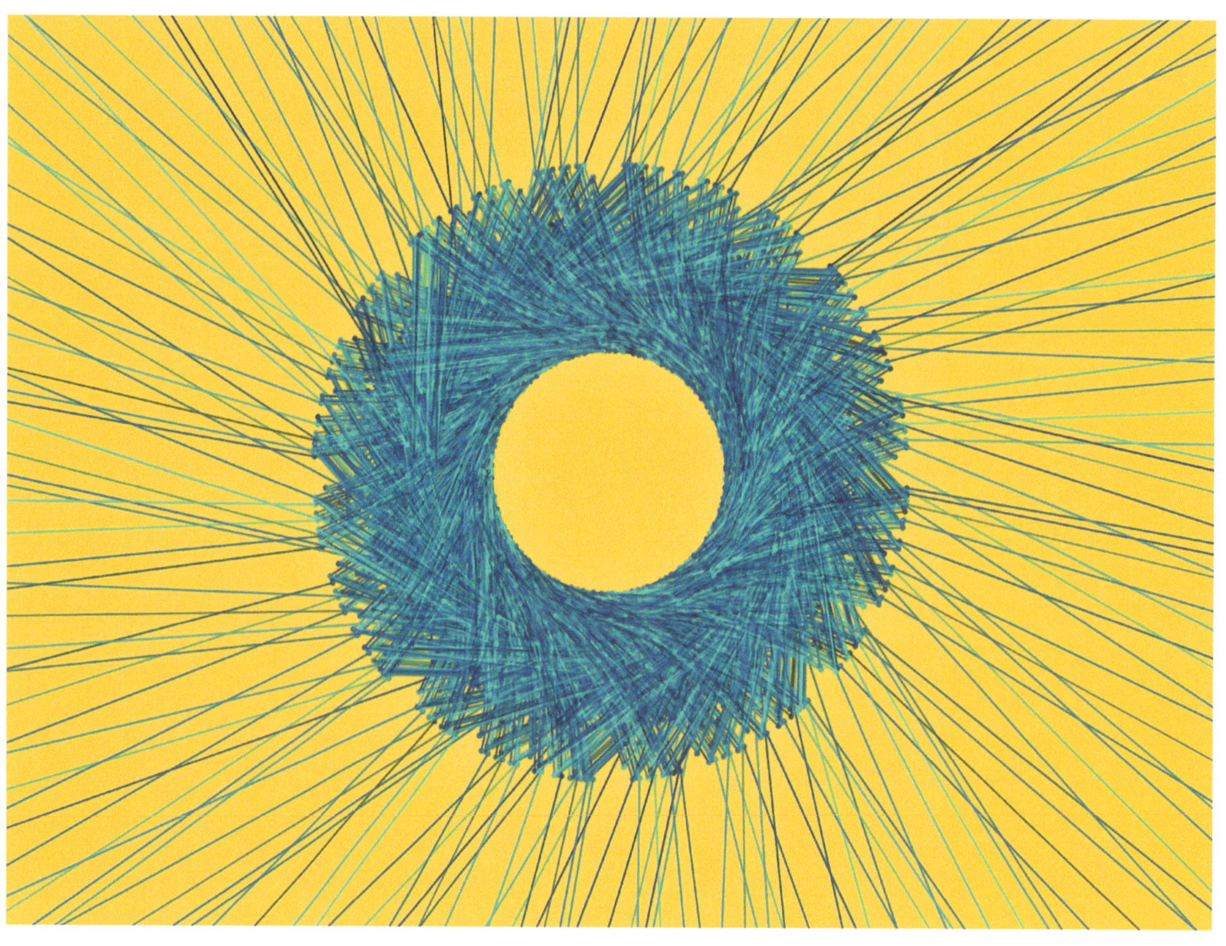

Figure 4.55: Traumwelt (Dreamworld)

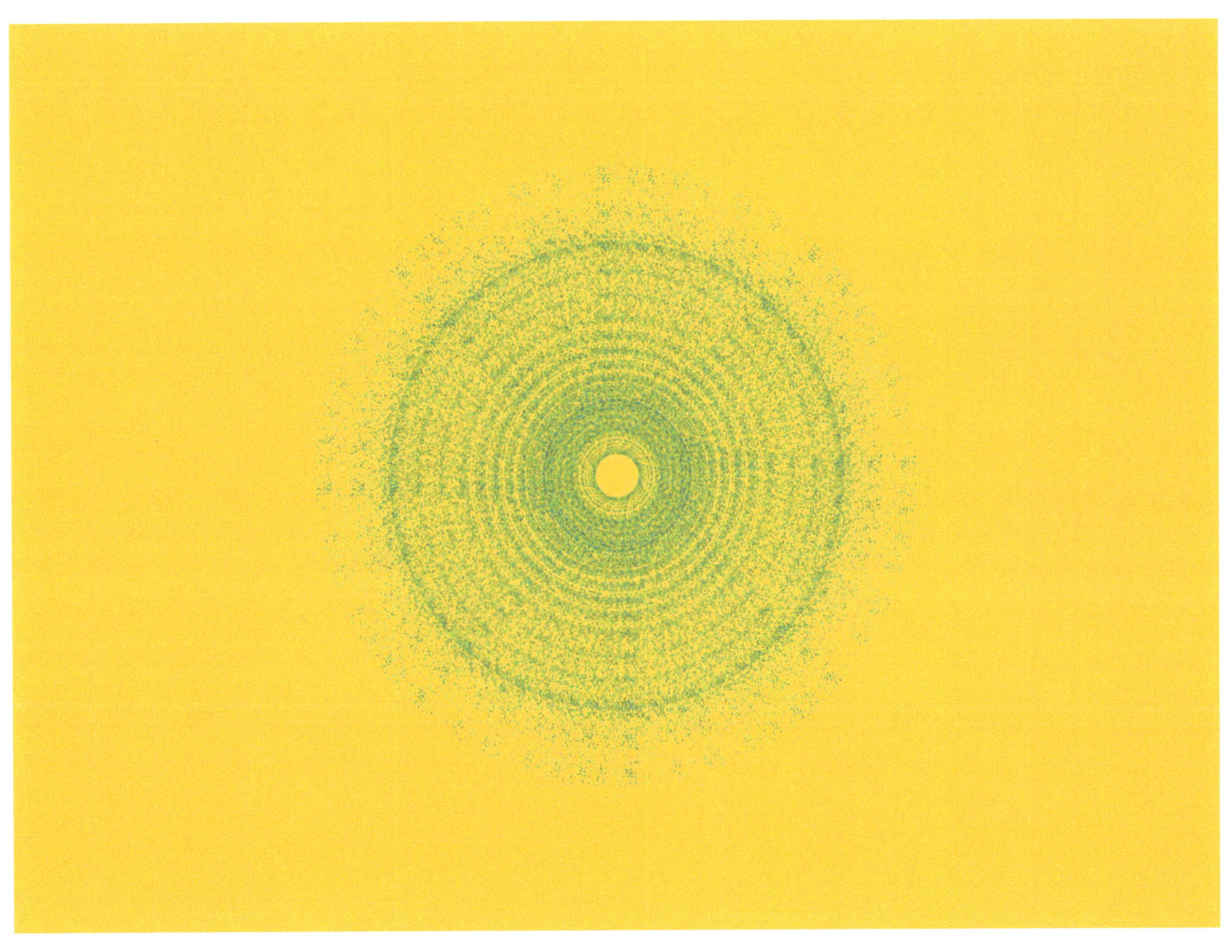

Figure 4.56: صمت (Silence)

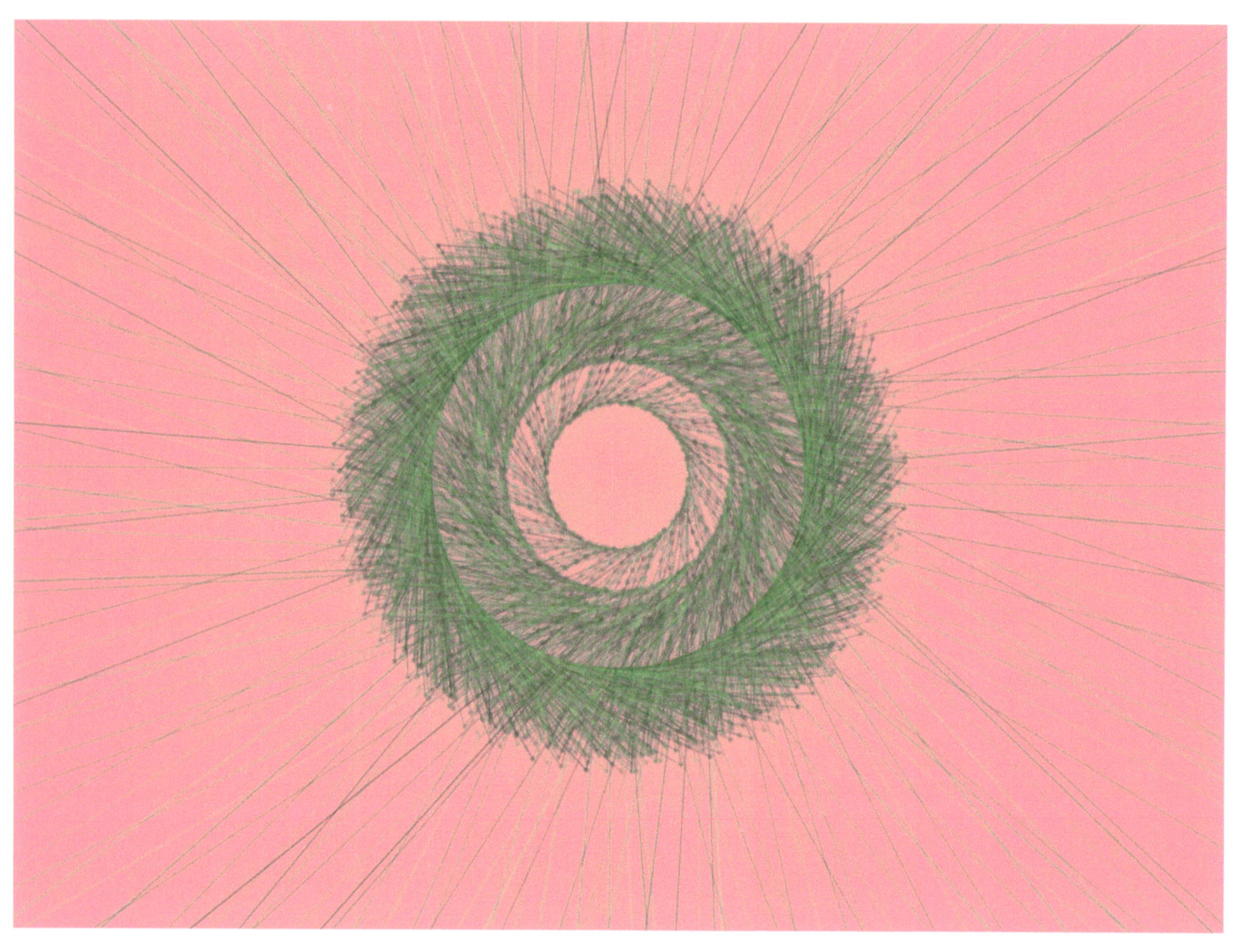

Figure 4.57: Contraste (Contrast)

Chapter 5

Animations with Generative Designer

An exciting aspect of generative art is the ability to create animations that showcase the evolution of a design over time.

There are two principal ways in which you can create animations with Generative Designer. The first method is the built-in "Animate" functionality that can be used by simply clicking the Animate button in the top right-hand corner of the menu. While this is a simple way to create animations, you cannot create animations with more than 20 frames and the output formats are limited to .webp. Further, the images that are created with this option are random instances of the same image input parameters.

But what if you want to create more sophisticated animations where you can piece together individual images of your own choosing? What if you want to combine images of different styles together in a long animation?

5.1 Using ffmpeg for Custom Animations

Luckily, you can do all that. Although creating such complex, custom animations is still a bit of a manual process, it is nonetheless easy to achieve. The technique involves using Generative Designer in conjunction with a free tool called ffmpeg.

Here is how to build your first animation.

First, create a series of images with the same background and style using Generative

Designer. To do this, simply adjust the settings and run the algorithm multiple times, saving each image with a different filename.

I suggest that you use the same background and style settings so there is an overall consistency in the type of image produced. Of course, you can choose to try this with more variable style settings, but I have found that the animations produced as a consequence seem more like slide shows than a smooth flowing animation.

Once you have created the images you wish to animate and have saved them to your disk, open the folder where the images are saved and ensure that the files are named sequentially, such as "image01.png", "image02.png", etc. If they are not named sequentially, you can use the command prompt to rename them. For example, to rename all files with the prefix "download" to "download01", "download02", etc., you can use the following command:

```
for %a in (download*.png) do rename "%a" "download%%02d.png"
```

This will rename all files with the prefix "download" to "download01.png", "download02.png", etc.

This step is necessary because ffmpeg expects to combine these images in an order and it will use these filenames and the embedded numbers to determine the order of frames.

Once the files are sequentially named, you can use ffmpeg to create an animation. If you don't already have ffmpeg installed, you can download it from the official website at `https://ffmpeg.org/`.

To create an animation, open a command prompt in the folder where the images are saved and execute the following command at the command prompt:

```
ffmpeg -r 4 -i image%02d.png -c:v libx264 -vf "scale=800:600" -crf 18 -
    ↪ pix_fmt yuv420p output.mp4
```

This command will create a video file called "output.mp4" from the images in the folder, with a frame rate of 4 frames per second and a resolution of 800x600 pixels. You can set different resolutions, for example, in order to create a 1080p quality output you would set the resolution to 1920x1080. The greater the resolution, the larger your animated video will be.

You can also adjust the parameters to suit your needs. For example, to create a GIF instead of a video, you can change the output file extension to ".gif". To change the frame rate, adjust the "-r" parameter. To change the resolution, adjust the "scale" parameter.

With Generative Designer and ffmpeg, beautiful and mesmerizing animations are simple to create.

Chapter 6

Future Directions

As small and simple as it is, Generative Designer is a powerful tool with immense potential for growth and improvement. How long would it take one of us to produce the art Generative Designer produces in a second? How quickly could we come up with the ideas it exposes us to, unaided? It is in human-machine collaboration that the power of generative design really lies. Generative tools can help us explore vast swathes of the realm of ideas and then tune into those that appeal to us and reflect our own goals.

Generative Designer will hopefully keep improving with time. What follows are some possible future directions for the development of the application, which can enhance the artistic capacity of users, make the application simpler to use and extend the functionality and range of outcomes possible with the program.

6.1 Control over Transparency of Lines

One possible enhancement to the Generative Designer application is the ability to control the transparency of lines with even greater control. Currently, the nodes at both ends of a line are three times wider than the line they connect and there is a single transparency control for both the node and the line. By allowing users to adjust node and line transparency separately, they can create more intricate and visually compelling designs, with overlapping layers that produce a greater sense of depth and complexity in the Generated art.

6.2   Separation of Connected Line Thickness and Node Thickness

Another possible improvement is the separation of connected line thickness and node thickness. This would give users even more control over the visual elements in their Generated art, enabling them to fine-tune their designs and create unique, visually appealing compositions by manipulating the thickness of individual elements.

6.3 Addition of Other Algorithms Beyond "Serendipitous Circles"

The inclusion of additional algorithms beyond the existing "Serendipitous Circles" would greatly enhance the versatility of the Generative Designer tool. By incorporating different algorithms, users can create a wider variety of art pieces, exploring new styles and techniques in Generative art. These algorithms could include, for example, fractals, geometric shapes, or other mathematically-inspired patterns.

6.4 Ability to Create Custom Animations with Greater Control

The Generative Designer application can be further enhanced by enabling users to create custom animations. The current animation capability is simple to use but limited to variations of the same type of image and to a maximum of twenty frames. In the future, Generative Designer could support a greater degree of custom control on the ordering of images, the number of images the system can generate and the addition of audio and other effects to the rendered video. This feature would open up new possibilities for artistic expression and allow users to create captivating, motion-filled art pieces.

6.5 Finer Grained Control Over Color

Providing users with finer grained control over color in their Generative art would be another valuable addition to the application. This could be achieved by allowing users to adjust individual color components (red, green, and blue) or by providing a more advanced color picker tool. With greater control over color, users can create more personalized and visually striking designs that better reflect their artistic vision.

6.6 Addition of Many New Color Palettes

Lastly, the addition of many new color palettes to the Generative Designer tool would further expand the creative possibilities available to users. By offering a wide range of color schemes, users can experiment with different aesthetics and moods in their generative art, discovering new styles and combinations that inspire their creativity. This enhancement would make the Generative Designer application an even more versatile and powerful tool for artistic expression.

Chapter 7

Explainable Art Algorithms vs Neural Networks

7.1 Introduction

Generative art has witnessed a surge in popularity in recent years, fueled by advances in artificial intelligence and machine learning. Two primary approaches to generative art have emerged: explainable art algorithms and neural networks. While both methodologies are capable of producing captivating and unique art, they employ vastly different techniques and cater to different artistic goals. In this chapter, we will compare these two approaches, delving into their strengths, weaknesses, and use cases, offering insights into their applications in the world of generative art.

7.2   Neural Networks for Generative Art

Neural networks, particularly deep learning models, have shown great promise in generating intricate and detailed art. Models such as DALL-E and MidJourney have gained significant attention for their ability to generate photo-realistic and imaginative images.

7.2.1 Advantages of Neural Networks

1. **Flexibility**: Neural networks are versatile and can generate a vast array of images, ranging from photo-realistic to highly stylized, depending on the training data.

2. **Complexity**: These models can produce images with rich details and intricate compositions, often surpassing human capabilities in terms of complexity.

3. **Adaptability**: Neural networks can be fine-tuned to cater to specific artistic styles or themes by adjusting the training data or network architecture.

7.2.2 Disadvantages of Neural Networks

1. **Resource Intensive**: Training neural networks for generative art requires vast amounts of computing power and data storage, often necessitating significant financial investment.

2. **Opaque Process**: Neural networks are inherently complex and often considered "black boxes." It is challenging to understand how the models generate images, and thus control over the final output can be limited.

3. **Accessibility**: Developing and training neural networks for generative art requires specialized knowledge in machine learning and programming, creating a barrier to entry for many aspiring artists.

7.3   Explainable Art Algorithms

Explainable art algorithms, such as those employed in Generative Designer, use mathematical functions and computational techniques to create abstract art. These algorithms offer a more accessible and transparent approach to Generative art.

7.3.1 Advantages of Explainable Art Algorithms

1. **Transparency**: The algorithms are easily understandable and can be modified by artists to produce desired effects, enabling greater control over the artistic process.

2. **Lightweight**: These algorithms are computationally efficient and can be executed on consumer-grade hardware, making them more accessible to a wider audience.

3. **Customizability**: Explainable art algorithms can be adapted and combined to create a wide variety of visual styles and effects, catering to different artistic tastes and preferences.

7.3.2 Disadvantages of Explainable Art Algorithms

1. **Limited Complexity**: These algorithms generally produce abstract art and are incapable of generating photo-realistic images or specific objects.

2. **Manual Fine-Tuning**: Artists must carefully adjust parameters and settings to achieve desired results, which can be time-consuming and require trial-and-error experimentation.

7.4 Combining Approaches for Enhanced Creativity

Artists can leverage the unique strengths of both neural networks and explainable art algorithms to create innovative and diverse works of art. By combining the complex, detailed imagery generated by neural networks with the more abstract and customizable output of explainable art algorithms, artists can unlock new creative possibilities and push the boundaries of generative art.

7.4.1 Hybrid Art Techniques

Several hybrid techniques can be employed to take advantage of both neural networks and explainable art algorithms, such as:

1. **Style Transfer**: Apply the visual style of one image, often generated by an explainable art algorithm, to the content of another image, often generated by a neural network. This technique allows artists to create unique combinations of styles and content.

2. **Layered Compositions**: Combine multiple layers of images generated by both neural networks and explainable art algorithms, using digital blending techniques to create visually rich and complex artworks.

3. **Post-processing Effects**: Apply digital filters, transformations, and other post-processing effects to the outputs of neural networks or explainable art algorithms, enhancing the visual impact of the generated images.

7.4.2 Collaborative Art Creation

Artists can collaborate with machine learning engineers and programmers to develop custom tools and algorithms that combine the power of neural networks with the accessibility and control offered by explainable art algorithms. This interdisciplinary approach can lead to the development of innovative and unique generative art techniques.

7.5 In Closing

Both neural networks and explainable art algorithms offer distinct advantages and limitations in the realm of generative art. While neural networks can generate highly detailed and complex images, their resource-intensive nature and inherent opacity can be limiting for many artists. In contrast, explainable art algorithms are lightweight, transparent, and easily customizable, offering a more accessible approach to generative art.

By understanding the strengths and weaknesses of both methodologies, artists can make informed decisions about which approach best suits their artistic goals and aspirations. Furthermore, the combination of these techniques can open up new creative possibilities, enabling artists to create entirely new forms of Generative art that push the boundaries of traditional artistic expression.

Chapter 8

Turning Designs into Objects

Generative art has the potential to go beyond the digital realm and manifest itself in the form of tangible, physical objects. Combining generative art with computer-aided design (CAD) programs and 3D printers enables artists and designers to create unique and innovative products that were previously impossible to produce using traditional manufacturing techniques. This chapter will explore how generative art can be integrated with CAD programs and 3D printers to create physical objects, discussing the use of generative art as both texture and templates for shape extrusion.

8.1 Generative Art and CAD Programs

CAD programs provide tools for designing and manipulating 3D models, which can then be exported for 3D printing or other manufacturing processes. By incorporating generative art into the CAD design process, artists can create complex and intricate designs that are difficult or impossible to achieve through manual design techniques.

There are several ways to incorporate generative art into the CAD design process:

- **Textures**: Generative art can be used to create textures that are applied to the surface of 3D models. These textures can add visual interest and complexity to otherwise simple objects.

- **Templates**: Generative art can also be used as a template for extruding shapes in the CAD environment. This allows for the creation of complex, organic forms that can be difficult to model using traditional CAD techniques.

One example of a free online CAD program that can be used to map generated art onto a 3D block for shape extrusion is TinkerCAD. Figure 8.1 shows an example of how TinkerCAD can be used for this purpose.

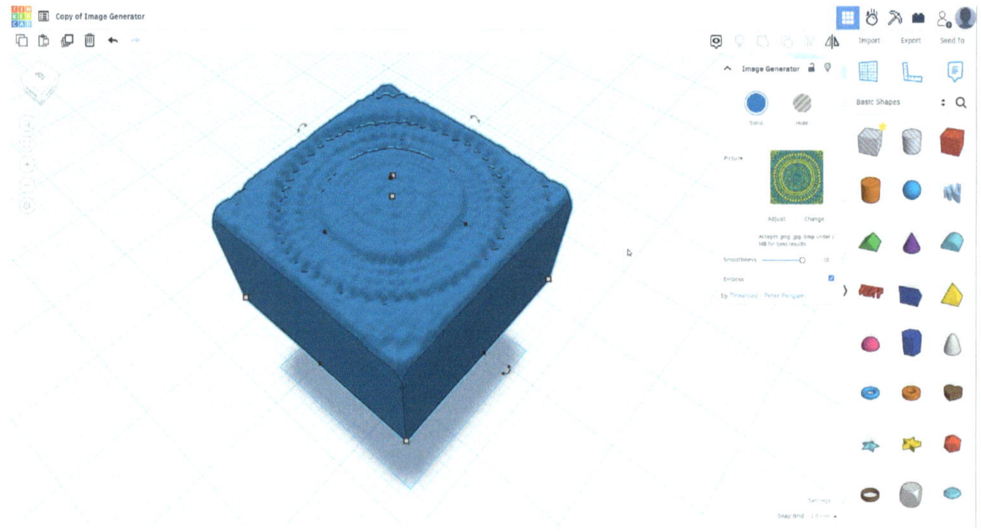

Figure 8.1: TinkerCAD example: Mapping generated art onto a 3D block for shape extrusion.

8.2   3D Printing Generative Art

Once a generative design has been incorporated into a CAD model, it can be exported as a 3D printable file format (such as STL or OBJ) and sent to a 3D printer for fabrication. The 3D printer will build the object layer by layer, following the contours of the design.

There are several factors to consider when 3D printing generative art:

- **Material selection**: The choice of material will affect the final appearance and physical properties of the printed object. Consider the desired look and feel of the object, as well as any functional requirements, when selecting a material.

- **Printing resolution**: The level of detail achievable in a 3D print is determined by the printer's resolution. Higher resolutions will produce smoother surfaces and finer details but may take longer to print.

- **Support structures**: Complex generative designs may require support structures to be printed alongside the object to ensure that it maintains its shape during the printing process. These structures will need to be removed after printing.

8.3 In Closing

Generative art can be used to create unique and innovative physical objects by combining its creative capabilities with CAD programs and 3D printing technology. By using generative art as both texture and templates for shape extrusion, artists can design intricate and complex forms that can then be realized as tangible objects. The fusion of these technologies expands the creative possibilities for artists and designers, allowing them to explore new avenues of artistic expression in the physical world.

Chapter 9

Further Reading and Resources

To deepen your understanding of generative art, explore various techniques and applications, and stay inspired, the following resources are highly recommended. They cover a range of topics related to generative design, p5.js, and the history of Serendipitous Circles.

9.1 Books

- Gross, B., Bohnacker, H., Laub, J., & Lazzeroni, C. (2018). *Generative Design: Visualize, Program, and Create with JavaScript in p5.js.* Princeton Architectural Press.

- McCarthy, L., Reas, C., & Fry, B. (2015). *Getting Started with p5.js: Making Interactive Graphics in JavaScript and Processing* (1st ed.). Make: Technology on Your Time.

- Arslan, E. (2018). *Learn JavaScript with p5.js: Coding for Visual Learners* (1st ed.). Apress.

- Fakhruddin, N. (n.d.). *Fun Art Projects in P5.js: For Beginners.* Kindle Edition.

9.2   Articles

- Kellerman, (1978). Serendipitous Circles Explored. *BYTE Magazine*, April 1978.

9.3 Videos

- Serendipitous Circles [Video]. (n.d.). Available from `youtube.com/watch?v=hhXB3_mHJQ4`

9.4 Tools

- ColorMind color palette generator [Website]. (n.d.). Available from `http://colormind.io/`

- p5.js multimedia library [Website]. (n.d.). Available from `https://p5js.org/download/`

- ZimJs Online Generative Art Tool [Website]. (n.d.). Available from `https://zimjs.com/cat/generator.html`

These resources provide good introductions and valuable insights into various aspects of generative art, programming techniques, and creative applications of the art you produce. They will help you become better equipped to create your own unique generative designs, experiment with different artistic approaches, modify Generative Designer or write your own programs and combine generative images with other forms of media, such as sound.

About the Author

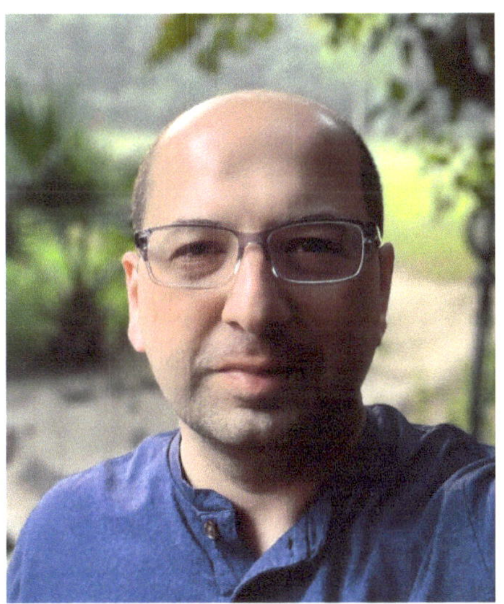

Amir Husain

Amir Husain is a serial entrepreneur, inventor, and author based in Austin, Texas. He has been named Austin's Top Technology Entrepreneur of the Year, listed as an Onalytica Top 100 Artificial Intelligence Influencer, and received the Austin Under 40 Technology and Science Award. Amir serves on the Board of Advisors for The University of Texas at Austin Department of Computer Science and on the NATO Maritime Unmanned Systems Innovation Advisory Board, and is a member of the Council on Foreign Relations. In 2013, Amir incorporated SparkCognition, an artificial intelligence company, where he has served as the Founder and CEO since its inception. Since its founding, SparkCognition has gained unicorn status and received widespread recognition. Amir is the author of the best-selling book, "The Sentient Machine," "Generative AI for Leaders," and the compilation, "Hyperwar." Amir is the Founding CEO of SkyGrid, a Boeing and SparkCognition company, and the Founding Chairman of Navigate, a decentralized data company. More information on the author and his work is available at: http://www.amirhusain.com